UCLA Symposia on Molecular and Cellular Biology, New Series

Series Editor, C. Fred Fox

RECENT TITLES

Volume 108
Acute Lymphoblastic Leukemia, Robert Peter Gale and Dieter Hoelzer, *Editors*

Volume 109
Frontiers of NMR in Molecular Biology, David Live, Ian M. Armitage, and Dinshaw J. Patel, *Editors*

Volume 110
Protein and Pharmaceutical Engineering, Charles S. Craik, Robert J. Fletterick, C. Robert Matthews, and James A. Wells, *Editors*

Volume 111
Glycobiology, Joseph K. Welply and Ernest Jaworski, *Editors*

Volume 112
New Directions in Biological Control: Alternatives for Suppressing Agricultural Pests and Diseases, Ralph R. Baker and Peter E. Dunn, *Editors*

Volume 113
Immunogenicity, Charles A. Janeway, Jr., Jonathan Sprent, and Eli Sercarz, *Editors*

Volume 114
Genetic Mechanisms in Carcinogenesis and Tumor Progression, Curtis Harris and Lance A. Liotta, *Editors*

Volume 115
Growth Regulation of Cancer II, Marc E. Lippman and Robert B. Dickson, *Editors*

Volume 116
Transgenic Models in Medicine and Agriculture, Robert B. Church, *Editor*

Volume 117
Early Embryo Development and Paracrine Relationships, Susan Heyner and Lynn M. Wiley, *Editors*

Volume 118
Cellular and Molecular Biology of Normal and Abnormal Erythroid Membranes, Carl M. Cohen and Jiri Palek, *Editors*

Volume 119
Human Retroviruses, Jerome E. Groopman, Irvin S.Y. Chen, Myron Essex, and Robin A. Weiss, *Editors*

Volume 120
Hematopoiesis, Steven C. Clark and David W. Golde, *Editors*

Volume 121
Defense Molecules, John J. Marchalonis and Carol L. Reinisch, *Editors*

Volume 122
Molecular Evolution, Michael T. Clegg and Stephen J. O'Brien, *Editors*

Volume 123
Molecular Biology of Aging, Caleb E. Finch and Thomas E. Johnson, *Editors*

Volume 124
Papillomaviruses, Peter M. Howley and Thomas R. Broker, *Editors*

Volume 125
Developmental Biology, Eric H. Davidson, Joan V. Ruderman, and James W. Posakony, *Editors*

Volume 126
Biotechnology and Human Genetic Predisposition to Disease, Charles R. Cantor, C. Thomas Caskey, Leroy E. Hood, Daphne Kamely, and Gilbert S. Omenn, *Editors*

Volume 127
Molecular Mechanisms in DNA Replication and Recombination, Charles C. Richardson and I. Robert Lehman, *Editors*

Volume 128
Nucleic Acid Methylation, Gary A. Clawson, Dawn B. Willis, Arthur Weissbach, and Peter A. Jones, *Editors*

Volume 129
Plant Gene Transfer, Christopher J. Lamb and Roger N. Beachy, *Editors*

Volume 130
Parasites: Molecular Biology, Drug and Vaccine Design, Nina M. Agabian and Anthony Cerami, *Editors*

Volume 131
Molecular Biology of the Cardiovascular System, Robert Roberts and Joseph F. Sambrook, *Editors*

Volume 132
Obesity: Towards a Molecular Approach, George A. Bray, Daniel Ricquier, and Bruce M. Spiegelman, *Editors*

Volume 133
Structural and Organizational Aspects of Metabolic Regulation, Paul A. Srere, Mary Ellen Jones, and Christopher K. Mathews, *Editors*

Please contact the publisher for information about previous titles in this series.

UCLA Symposia Board

C. Fred Fox, Ph.D., Director
Professor of Microbiology, University of California, Los Angeles

Charles J. Arntzen, Ph.D.
Director, Plant Science and Microbiology
E.I. du Pont de Nemours and Company

Floyd E. Bloom, M.D.
Director, Preclinical Neurosciences/
Endocrinology
Scripps Clinic and Research Institute

Ralph A. Bradshaw, Ph.D.
Chairman, Department of Biological
Chemistry
University of California, Irvine

Francis J. Bullock, M.D.
Vice President, Research
Schering Corporation

Ronald E. Cape, Ph.D., M.B.A.
Chairman
Cetus Corporation

Ralph E. Christoffersen, Ph.D.
Executive Director of Biotechnology
Upjohn Company

John Cole, Ph.D.
Vice President of Research
and Development
Triton Biosciences

Pedro Cuatrecasas, M.D.
Vice President of Research
Glaxo, Inc.

Mark M. Davis, Ph.D.
Department of Medical Microbiology
Stanford University

J. Eugene Fox, Ph.D.
Vice President, Research
and Development
Miles Laboratories

J. Lawrence Fox, Ph.D.
Vice President, Biotechnology Research
Abbott Laboratories

L. Patrick Gage, Ph.D.
Director of Exploratory Research
Hoffmann-La Roche, Inc.

Gideon Goldstein, M.D., Ph.D.
Vice President, Immunology
Ortho Pharmaceutical Corp.

Ernest G. Jaworski, Ph.D.
Director of Biological Sciences
Monsanto Corp.

Irving S. Johnson, Ph.D.
Vice President of Research
Lilly Research Laboratories

Paul A. Marks, M.D.
President
Sloan-Kettering Memorial Institute

David W. Martin, Jr., M.D.
Vice President of Research
Genentech, Inc.

Hugh O. McDevitt, M.D.
Professor of Medical Microbiology
Stanford University School of Medicine

Dale L. Oxender, Ph.D.
Director, Center for Molecular Genetics
University of Michigan

Mark L. Pearson, Ph.D.
Director of Molecular Biology
E.I. du Pont de Nemours and Company

George Poste, Ph.D.
Vice President and Director of Research and
Development
Smith, Kline and French Laboratories

William Rutter, Ph.D.
Director, Hormone Research Institute
University of California, San Francisco

George A. Somkuti, Ph.D.
Eastern Regional Research Center
USDA-ARS

Donald F. Steiner, M.D.
Professor of Biochemistry
University of Chicago

UCLA Symposia Board membership at the time of the meeting is indicated on the above list.

Early Embryo Development and Paracrine Relationships

Early Embryo Development and Paracrine Relationships

Proceedings of a UCLA Symposia Colloquium,
Held at Taos, New Mexico
February 3-8, 1989

Editors

Susan Heyner

Obstetrics and Gynecology Research Laboratory
Albert Einstein Medical Center
Philadelphia, Pennsylvania

Lynn M. Wiley

California Primate Research Center
University of California
Davis, California

A JOHN WILEY & SONS, INC., PUBLICATION
New York • Chichester • Brisbane • Toronto • Singapore

Address all Inquiries to the Publisher
Alan R. Liss, Inc., 41 East 11th Street, New York, NY 10003

Copyright © 1990 Alan R. Liss, Inc.

Printed in United States of America

Under the conditions stated below the owner of copyright for this book hereby grants permission to users to make photocopy reproductions of any part or all of its contents for personal or internal organizational use, or for personal or internal use of specific clients. This consent is given on the condition that the copier pay the stated per-copy fee through the Copyright Clearance Center, Incorporated, 27 Congress Street, Salem, MA 01970, as listed in the most current issue of "Permissions to Photocopy" (Publisher's Fee List, distributed by CCC, Inc.), for copying beyond that permitted by sections 107 or 108 of the US Copyright Law. This consent does not extend to other kinds of copying, such as copying for general distribution, for advertising or promotional purposes, for creating new collective works, or for resale.

The publication of this volume was facilitated by the authors and editors who submitted the text in a form suitable for direct reproduction without subsequent editing or proofreading by the publisher.

Library of Congress Cataloging-in-Publication Data

Early embryo development and paracrine relationships : proceedings of a UCLA colloquium held at Taos, New Mexico, February 3-8, 1989 / editors, Susan Heyner, Lynn M. Wiley.
    p.   cm. -- (UCLA symposia on molecular and cellular biology ; new ser., v. 117)
  Includes bibliographical references.
  ISBN 0-471-56753-1
  1. Embryology--Congresses. 2. Ovum implantation--Congresses.
3. Growth factors--Physiological effect--Congresses. 4. Molecular biology--Congresses.   I. Heyner, Susan. II. Wiley, Lynn M.
III. Series.
QP277.E2   1990
599'.033--dc20   89-70598
                                                                  CIP

Contents

Contributors	ix
Preface	
Susan Heyner and Lynn M. Wiley	xi
Bovine Oocyte Maturation and Embryo Development	
W.H. Eyestone and N.L. First	1
The Expression of Growth Factor Ligands and Receptors in Preimplantation Mouse Embryos	
Daniel A. Rappolee, Karin S. Sturm, Gilbert A. Schultz, Roger A. Pedersen, and Zena Werb	11
Changes in RNA and Protein Synthesis During Development of the Preimplantation Mouse Embryo	
Gilbert Schultz, Wendy Dean, Ann Hahnel, Nancy Telford, Daniel Rappolee, Zena Werb, and Roger Pedersen	27
Detection of EGF Receptor Protein and Activities in Embryo Cells	
Eileen D. Adamson and Mark Mercola	47
Developmental Regulation of the KS-FGF Oncogene by Embryonal Carcinoma Cells and Early Mouse Embryos	
Angie Rizzino, Jay Tiesman, Ronald Hines, and David Kelly	53
The Energy Metabolism of the Preimplantation Embryo	
Henry J. Leese	67
Regulation of Hamster Preimplantation Embryo Development In Vitro by Glucose and Phosphate	
Barry D. Bavister	79
Gene Expression Required for Blastocoel Formation in the Mouse	
Gerald M. Kidder and Andrew J. Watson	97
The Role of Insulin in Preimplantation Mouse Development	
L.V. Rao, M. Farber, R.M. Smith, and S. Heyner	109
Epidermal Growth Factor and Pregnancy in the Mouse	
Y.M. Huet-Hudson, G.K. Andrews, and S.K. Dey	125
Insulin-Like Growth Factor Binding Protein and Pregnancy: Regulation and Function in the Primate	
A.T. Fazleabas, H.G. Verhage, and S.C. Bell	137
***In Vitro* Implantation on Polarized Uterine Epithelia**	
Stanley R. Glasser, Joanne Julian, Joy Mulholland, Shailaja Mani, Daniel D. Carson, and Andrew L. Jacobs	153
Index	169

Contributors

Eileen D. Adamson, La Jolla Cancer Research Foundation, La Jolla, CA 92037 **[47]**

G.K. Andrews, Department of Biochemistry and Molecular Biology, Ralph L. Smith Research Center, University of Kansas Medical Center, Kansas City, KS 66103 **[125]**

Barry D. Bavister, Department of Veterinary Science, University of Wisconsin, Madison, WI 53706 **[79]**

S.C. Bell, Departments of Obstetrics and Gynecology and Biochemistry, University of Leicester, Leicester LE2 7LX, United Kingdom **[137]**

Daniel D. Carson, Department of Biochemistry and Molecular Biology, M.D. Anderson Cancer Institute, Houston, TX 77030 **[153]**

Wendy Dean, Department of Medical Biochemistry, University of Calgary, Alberta, Canada T2N 4N1 **[27]**

S.K. Dey, Departments of Obstetrics–Gynecology and Physiology, Ralph L. Smith Research Center, University of Kansas Medical Center, Kansas City, KS 66103 **[125]**

W.H. Eyestone, Department of Meat and Animal Science, University of Wisconsin, Madison, WI 53706 **[1]**

M. Farber, Department of Obstetrics and Gynecology, Albert Einstein Medical Center, Philadelphia, PA 19141 **[109]**

A.T. Fazleabas, Department of Obstetrics and Gynecology, University of Illinois, Chicago, IL 60612 **[137]**

N.L. First, Department of Meat and Animal Science, University of Wisconsin, Madison, WI 53706 **[1]**

Stanley R. Glasser, Department of Cell Biology, Baylor College of Medicine, Houston, TX 77030 **[153]**

Ann Hahnel, Department of Medical Biochemistry, University of Calgary, Alberta, Canada T2N 4N1 **[27]**

S. Heyner, Department of Obstetrics and Gynecology, Albert Einstein Medical Center, Philadelphia, PA 19141 **[109]**

Ronald Hines, Eppley Institute, University of Nebraska Medical Center, Omaha, NE 68105-1065 **[53]**

The numbers in brackets are the opening page numbers of the contributors' articles.

Contributors

Y.M. Huet-Hudson, Department of Physiology, Ralph L. Smith Research Center, University of Kansas Medical Center, Kansas City, KS 66103; present address: Monsanto Company, St. Louis, MO 63198 [125]

Andrew L. Jacobs, Department of Biochemistry and Molecular Biology, M.D. Anderson Cancer Institute, Houston, TX 77030 [153]

Joanne Julian, Department of Cell Biology, Baylor College of Medicine, Houston, TX 77030 [153]

David Kelly, Eppley Institute, University of Nebraska Medical Center, Omaha, NE 68105-1065 [53]

Gerald M. Kidder, Department of Zoology, University of Western Ontario, London, Canada N6A 5B7 [97]

Henry J. Leese, Department of Biology, University of York, Heslington, Y01 5DD, United Kingdom [67]

Shailaja Mani, Department of Cell Biology, Baylor College of Medicine, Houston, TX 77030 [153]

Mark Mercola, Dana Farber Cancer Institute, Harvard Medical School, Boston, MA 02115 [47]

Joy Mulholland, Department of Cell Biology, Baylor College of Medicine, Houston, TX 77030 [153]

Roger Pedersen, Laboratory of Radiobiology and Environmental Health and the Department of Anatomy, University of California, San Francisco, CA 94143 [11, 27]

L.V. Rao, Department of Obstetrics and Gynecology, Albert Einstein Medical Center, Philadelphia, PA 19141 [109]

Daniel A. Rappolee, Laboratory of Radiobiology and Environmental Health and the Department of Anatomy, University of California, San Francisco, CA 94143 [11, 27]

Angie Rizzino, Eppley Institute, University of Nebraska Medical Center, Omaha, NE 68105-1065 [53]

Gilbert A. Schultz, Department of Medical Biochemistry, University of Calgary, Alberta, Canada T2N 4N1 [11, 27]

R.M. Smith, Department of Pathology and Laboratory Medicine, University of Pennsylvania, Philadephia, PA 19104 [109]

Karin S. Sturm, Laboratory of Radiobiology and Environmental Health and the Department of Anatomy, University of California, San Francisco, CA 94143 [11]

Nancy Telford, Department of Medical Biochemistry, University of Calgary, Alberta, Canada T2N 4N1 [27]

Jay Tiesman, Eppley Institute, University of Nebraska Medical Center, Omaha, NE 68105-1065 [53]

H.G. Verhage, Department of Obstetrics and Gynecology, University of Illinois, Chicago, IL 60612 [137]

Andrew J. Watson, Department of Zoology, University of Western Ontario, London, Canada N6A 5B7 [97]

Zena Werb, Laboratory of Radiobiology and Environmental Health and the Department of Anatomy, University of California, San Francisco, CA 94143 [11, 27]

Preface

The 1989 UCLA Colloquium on **Early Embryo Development and Paracrine Relationships** provided molecular and cellular developmental and reproductive biologists with an opportunity to review and discuss mechanisms underlying early mammalian growth and development and certain aspects of maternal–embryonic interaction. The first part of the meeting was devoted to the physiology of preimplantation embryos, with major emphasis on mechanisms of genetic regulation that underlie preimplantation development. Progress in the study of nutrient transport across the plasma membrane of the early embryo has substantially enhanced our understanding of early metabolic patterns and led to improvements in the design of media for embryo culture. The development of the reverse transcriptase–polymerase chain reaction (RT–PCR) has facilitated analysis of mRNA transcripts in extremely small amounts of tissue, as little as one mouse embryo. The use of this powerful technique has shown that mRNA transcripts for a number of growth factors, as well as their receptors, may be detected in preimplantation stages of development. Continued work along these lines may provide insights into those maternal–embryo interactions that sustain implantation and development, and those whose dysfunction leads to early pregnancy loss.

Since the embryo is multicellular, and it is well established that cell interactions underlie morphogenesis, there is significant interest in identifying molecular candidates that may mediate the transduction of cellular interactions into morphogenetic events. Two of these candidates are plasma membrane ion transporters that are known to respond to growth factors in the adult and that have a developmentally regulated, polar distribution on blastomeres. These ion transporters have been implicated in blastocoele formation and in the differentiation of the first two lineages to form in the embryo, trophectoderm and inner cell mass. It will be very exciting if future work can link the expression of growth factors to the induction and/or activity of these (or additional) transporters.

The second part of the colloquium was devoted to discussion of the signals that are involved in successful implantation and a consideration of the maternal component, the decidua and endometrium. Implantation is the most vulnerable phase of reproduction and is also far more inaccessible than the earlier phase of the free-living preimplantation embryo. The significance of signalling between the embryo and endometrium was re-emphasized, and the nature of such signals was discussed.

The colloquium generated considerable discussion, as well as laying the ground for future collaborative interactions.

We would like to acknowledge generous support from the Director's Sponsor Fund established by E.I. du Pont de Nemours and Company, Inc.; Hoffmann-La Roche, Inc.; Immunex, Inc.; Monsanto Corporation; Schering Corporation; and the Upjohn Company. The success of the meeting was due in large part to excellent organization by the UCLA Symposia staff. We would like to single out Dr. Robin Yeaton-Woo for special thanks, and also Jason Gursky for on-site arrangements, Hank Harwood for budget arrangements, and Betty Handy for editorial assistance. Finally, we wish to thank the speakers and participants for their timely presentations and stimulating discussions.

Lynn M. Wiley
Susan Heyner

BOVINE OOCYTE MATURATION AND EMBRYO DEVELOPMENT[1]

W.H. Eyestone and N.L. First

Department of Meat and Animal Science
University of Wisconsin
Madison, WI 53706

ABSTRACT Highly efficient systems for preparation of sperm for in vitro fertilization exist for most mammalian species studied. The principal limitations in producing embryos in vitro are limits in developmental competence of cultured oocytes and in embryo development in culture.

Fertilization and the first three to four cleavage divisions occur in the oviduct, after which the embryo passes into the uterus where compaction, blastulation and the balance of embryonic and fetal development occurs. Several landmark developmental events occur during this period, including the shift from meiosis to mitosis and the onset of embryonic gene expression. In many mammalian species, the study of early development is complicated by embryonic refractoriness to conventional in vitro culture methods, the rabbit and some inbred strains of mice being notable exceptions. In cattle embryos cleavage in vitro arrests at the 8- to 16-cell stage. The cause of this block is unknown. Although the connections are unclear, the block stage coincides with the onset of embryonic transcription and an abrupt increase in cell cycle length. In vitro blocks may be overcome by co-culturing embryos with oviductal tissue, or in medium conditioned by oviductal tissue, suggesting an embryonic dependence on tubal factors for normal cleavage.

---

[1]This research was supported by the College of Agricultural and Life Sciences, University of Wisconsin-Madison, a grant from W.R. Grace and Company and USDA grant #144 AM 75.

INTRODUCTION

The recent development of methods for producing bovine embryos through in vitro maturation, fertilization and culture has provided tools for studying the influences of the oocyte and fertilization on embryo development. It has also provided a means for timing and characterizing important events during early bovine development. This review will focus on problems of bovine embryo development associated with in vitro maturation of oocytes and on the physiological mechanisms behind those problems as well as methods for solution. It will also focus on the use of in vitro produced embryos to study bovine embryo development.

Oocyte Maturation.

As with oocytes of all mammalian species, bovine oocytes go through an initial period of mitotic proliferation followed by initiation of meiosis and meiotic dictyate arrest followed by a growth phase during which meiotic arrest continues, cytoplasmic volume expands and the zona pellucida is synthesized. The growth phase occurs primarily within the secondary and early tertiary stages of ovarian follicles and is the period when massive protein synthesis occurs. During the late oocyte growth RNAs are stored for translation later, primarily during the maternal period of embryonic development (1). Bovine oogonia can be identified as early as Day 25 of gestation and the first oogonia terminate mitosis and enter meiosis around Day 80 of gestation. The first oogonia begin dictyate arrest around Day 95 and by Day 130 essentially all oocytes are in dictyate arrest. There is an overlapping of the period of mitosis, initiation of meiosis and establishment of dictyate arrest. Bovine oocytes remain in dictyate arrest until stimulated by the ovulatory LH surge or until removed from a tertiary follicle.

The follicular mechanisms preventing meiotic resumption and oocyte maturation appear to be similar between the bovine and the mouse. For example, if oocytes are removed from follicles and immediately transferred to cultures containing granulosa cells, follicular fluid, hypoxanthine, adenosine, cyclic AMP or its analogs, meiotic arrest will be maintained (2). Oocyte maturation usually refers to the ability of an oocyte to progress from the germinal vesicle stage at follicle removal to second metaphase. For the purpose of this review, the term maturation will be broadened to include not only nuclear maturation, but cumulus maturation (expansion) and cytoplasmic maturation; i.e., acquisition of competence to complete fertilization and embryo development. In resumption of meiosis, the bovine oocyte first undergoes a period of commitment to germinal vesicle breakdown. This period has been shown to be less than 1 hr and recent evidence indicates it may be as short as 20 to 30 minutes (2). During this period agonists of the cyclic AMP second messenger system will inhibit meiotic resumption. If applied after the commitment period, follicular fluid, granulosa cells, or cyclic AMP are ineffective in preventing resumption of meiosis. RNA (3) and protein synthesis are required for the resumption of meiosis (2,4) and protein synthesis must occur within the first 3 hr of removal from the follicle. Discrete proteins are required for germinal vesicle breakdown, completion of metaphase I, completion of metaphase II, and for the arrest of the oocyte at metaphase II (2).

The expansion of cumulus cells occurs in vivo before ovulation and is stimulated by gonadotropins. In vitro expansion of the cumulus cells is also stimulated by the addition of gonadotropins and particularly FSH to the culture media (5). Bovine oocytes can be fertilized with or without cumulus cells; however, fertilization may be delayed when cumulus cells are present (Parrish, unpublished). Bovine cumulus cells are also known to produce a protein which stimulates sperm motility (6). It is likely that the principal function of cumulus expansion in the bovine is as in the hamster to provide a large sticky mass which can be picked up by the fimbria of the oviduct. The cumulus is short lived in the bovine and is normally not present on oocytes or embryos in the oviduct after 2-3 hr (7).

Cytoplasmic Maturation.

The acquisition by the cytoplasm of the ability to complete embryo development is dependent upon the size of follicle from which the oocyte is removed. This ability has been termed acquisition of developmental competence. In the bovine fertilization of oocytes recovered from follicles of 1-3 mm and cultured in the absence of cumulus cells results in normal oocyte maturation and fertilization but failed embryo development (8). This developmental incompetence can be corrected by co-culture of the oocyte during the in vitro maturation period with granulosa cells or with an equal and abundant supply (>1 x 10^6 cells/ml) of cumulus cells (9-11). Acquisition of developmental competence by oocytes from small follicles is dependent on the presence of granulosa or cumulus cells (12). Approximately 77-80% of the bovine follicles are considered to be atretic (13). To what extent oocytes from atretic follicles can be rescued in their developmental competence is unknown. However, oocytes showing signs of coming from severely atretic follicles, that is, oocytes that are missing a major portion of the cumulus and oocytes which have lost their cytoplasmic integrity, have very little developmental or fertilization competence. By screening oocytes for normal morphology to avoid those from atretic follicles, by controlling the number of oocytes per culture, controlling the culture conditions and by co-culturing with granulosa cells or abundant cumulus cells, developmental competence of oocytes from small follicles can be restored to nearly the level of developmental competence from oocytes of superovulated cattle matured in vivo. This developmental competence can be as high as 40 to 45% for oocytes properly matured in vitro and approximately 55% for oocytes matured in vivo (11). Embryo development in the bovine is also influenced by the sire supplying the semen to fertilize the oocyte. Eyestone (14) equalized fertilization across bulls by adjusting sperm concentrations to rule out differences due to fertilization. Nevertheless, the proportion of blastocysts obtained varied among bulls. Whether this is a genetic or epigenetic effect is unknown.

Bovine embryos, whether produced in vitro or in vivo, are generally blocked in culture between the 8-16 cell stage (14). They are blocked in response to exposure to culture conditions during the previous cell cycle, that is oocytes blocked at the 8-16 cell stage are blocked because they were exposed to in vitro culture at the 4-6 cell stage (14). Oocytes placed in culture at the 8-cell stage do not block. Bovine embryos can be cultured through the block by incubation in the oviducts of surrogate rabbits (reviewed by Boland, 15) or sheep (16). More recently bovine embryos have been co-cultured through the block with oviduct epithelial cells (17,18). Eyestone and First (18) have shown that bovine oocytes can be cultured from one-cell to blastocysts in conditioned media coming from the culture of oviduct epithelial cells. This conditioned media can be stored frozen (14).

The ability to produce bovine oocytes in vitro with precise synchrony has allowed study of the timing of bovine embryo development as well as the timing of components of each cell cycle throughout embryo development. It has also allowed determination of the time when the bovine embryo transitions from maternal to zygotic control of development and identification of the early proteins produced by the embryo.

Cleavage.

In vitro matured and fertilized ova provide an excellent system in which to study the chronology of early cleavage largely due to the fact that sperm penetration, and thus oocyte activation occurs synchronously in in vitro fertilized oocytes. For example, Barnes et al. (19) characterized the timing of oocyte penetration by sperm from five bulls during in vitro fertilization and found that while penetration times varied among bulls, the majority fertilized by a given bull were penetrated within a span of 2 hr. Synchronous fertilization yields populations of embryos which cleave synchronously and are thus suitable for intensive chronological studies.

Cell Cycle 1.

Barnes et al. (19) and Eyestone and First (20) have characterized the early cell cycles of in vitro matured, in vitro fertilized bovine embryos cultured in vitro. Cell cycle phases were determined after microfluorometric quantitation of nuclear DNA (20). The first cell cycle (1 cell) was timed in hours post-insemination. Subsequent cycles were timed in hours post-cleavage. Groups of embryos were synchronized to the preceding cleavage division by selecting newly-cleaved embryos at bi-hourly intervals. After insemination or cleavage, embryos were removed at 2 hr intervals for DNA analysis. Embryos were air-dried onto glass slides and fixed in 70% ethanol for 2 hr. Nuclear DNA was stained with DAPI according to the procedures of Hamada and Fujita (21). Since DAPI binds quantitatively to DNA, the fluorescence emitted by the DAPI-DNA complex in the presence of excess DAPI is proportional to the amount of DNA bound. DNA was quantitated by measuring fluorescence with the aid of computerized image analysis.

Mean fluorescence intensity per nucleus was calculated for each time point. Granulosa cell nuclei served as a diploid 2C standard and were prepared and analyzed in the same manner and within the same replicates as embryonic nuclei. Mean embryonic nuclear fluorescence at each time point was divided by the mean granulosa cell reading, providing a ratio of embryonic DNA to the 2C standard. These ratios were expressed as a function of time to plot cell cycle.

According to these studies, the time from fertilization to the completion of meiosis 2 was 1-2 hr; pronuclear formation was completed by 2 hr. The period between completion of meiosis and the onset of DNA synthesis was 6-7 hr (G_1). DNA synthesis lasted approximately 8 hr. The first cell cycle also included a rather long G_2/M phase which lasted about 6 hr. Metaphase chromosomes were observed 4-6 hr after the onset of G_2. Cytokinesis followed the appearance of metaphase chromosomes by about 2 hr.

Cell Cycle 2.

The second cell cycle has been analyzed in a manner similar to the first. Compared to cell cycle 1, cycle 2 is quite rapid, lasting only 12 to 15 hr, compared to 22 hr for cycle 1. This shortening is due primarily to the lack of a prominent G_1 phase, which lasts a maximum of 2 hr (if it exists at all). The major portion of cycle 2 is occupied by S (8 hr), followed by a 2 to 4 hr G_2/M phase. Metaphase chromosomes were seen to form within 2 hr of the completion of S.

Cycle 3.

Assessing the third cycle is complicated somewhat by asynchronous cleavage at both ends of the cell cycle. This situation raises the question of whether cycle length should be measured as "doubling time" of all blastomeres or as the intermitotic period of individual blastomeres. Barnes et al. (19) reported that the period between the appearance of 4 cells to 5 to 6 cells was 9 hr, and to 8 cells, 15 hr. These data reveal a rather high degree of asynchronous cleavage in which division among all four blastomeres occurs over a 6 hr period. These data also reveal that cycle times range from 9 to 15 hr among blastomeres, raising the possibility that at least two populations of cell cycle types develop at this stage.

Protein synthesis undergoes important modifications during the third cycle. In ovine embryos, total protein synthesis declines precipitously to 5% of earlier values, and remains low through the fourth or fifth cycle (22). Indirect evidence suggests that a similar decline occurs in protein synthesis during the third cell cycle of bovine embryos. Barnes (19) cultured embryos in the presence of ^{35}S-methionine at various stages between 1 and 16 cells, and visualized the newly synthesized proteins fluorographically after 2-D gel electrophoresis. An abrupt drop in fluorographic exposure density was observed between the early and late 4-cell stage. Since cells at each cleavage stage were made under identical conditions, this observation suggests, in a qualitative fashion, a marked drop in total protein synthesis at this time.

The third cell cycle also marks the point at which the embryonic genome is activated. Barnes (19) demonstrated the expression of a small group of α-amanitin sensitive proteins during the late 4-cell stage. The disappearance of several proteins at this stage was also inhibited by α-amanitin. These protein changes were coincident with the drop in total protein synthesis discussed in the preceding paragraph.

In addition to these internal changes, embryos exhibit a marked sensitivity to in vitro culture conditions during the third cycle. In an experiment designed to detect the existence of such stage-related sensitivity, 1- to 2-, 4-, 8- or 16-cell embryos were collected from donors cultured 24 hr in vitro, and then transferred to ligated ovine oviducts. Controls at each stage were transferred immediately to ovine oviducts. Development to blastocyst in oviducts was unaffected by in vitro culture of all stages except for the 4-cell stage, where development to blastocyst was reduced 50% by in vitro culture. These results indicate that embryos were more sensitive to in vitro conditions during the third cell cycle.

Cell Cycle 4.

The fourth cell cycle marks the onset of major changes in developmental pattern. For the first time, blastomeres become related to either an outer, subzonal location or a central location. Such localization may be important in light of the "inside-outside" hypothesis of cell fate; i.e., that outer blastomeres give rise to trophoblast and inner blastomeres to inner cell mass. The duration of the fourth cycle is also much longer than preceding cycles, lasting from 20-30 hr. Determination of the fourth cycle's duration has been complicated by the 8-16 cell block to in vitro development. The recent introduction of co-culture systems capable of alleviating this block (18) should facilitate analysis of this cell cycle.

One of the most striking events of the fourth cycle involves the appearance of a plethora of new proteins resulting from genomic activation. Details on the regulation of this event are beginning to emerge from studies on Drosophila and Xenopus embryos. In those species, genomic activation is temporally associated with abrupt increases in cell cycle length (due to the appearance of G_1 and/or G_2 phases) and abrupt decreases in protein synthesis. Indeed, in Xenopus embryos, the timing of genomic activation and the onset of lengthened cell cycles can be advanced by suppressing protein synthesis. Genomic activation in cattle (19), sheep (22) and mice (23) is associated with extended cell cycles. Furthermore, this event occurs after an abrupt drop in protein synthesis in cattle and sheep (22,24).

Superimposed on this sequence of events is the stage-specific susceptibility of bovine and ovine embryos to "block" in vitro. While such blocks most certainly result from inadequacies in currently used culture systems, the fact that such blocks occur at a time coincident with abrupt declines in protein synthesis and genomic activation raises some interesting questions. For example, do blocks in vitro reflect the failure of genomic activation? Do blocks in vitro reflect a culture system's inability to support protein synthesis between the late 4- to 8-cell stage above a threshold level necessary for cell function?

Our picture of cell cycle events during early cleavage is far from complete. We do, however, feel that approaching the regulation of early development as a special case of cell cycle regulation will allow us to tap the wider body of information on cell cycle regulation in other cell types. This approach is particularly valid in light of accumulating evidence that many important aspects of cell cycle regulation are common to all eukaryotes and may thus offer clues into the fundamentals of cell reproduction in all species and cell types.

REFERENCES

1. Bachvarova R, Payton BV (1988). Gene expression during growth and meiotic maturation of mouse oocytes. In Haseltine FP, First NL (eds): "Meiotic Inhibition: Molecular Control of Meiosis", New York: Alan R. Liss, p 67.
2. Sirard MA, First NL (1988). In vitro inhibition of oocyte nuclear maturation in the bovine. Biol Reprod 39:229.
3. Hunter AG, Moor RM (1986). In vitro blockage of germinal vesicle breakdown (GVBD) in bovine oocytes by RNA inhibitors. J Dairy Sci 69:145.
4. Hunter AG, Moor RM (1987). Stage-dependent effects of inhibiting ribonuclear acids and protein synthesis on meiotic maturation of bovine oocytes in vitro. J Dairy Sci 70:1646.
5. Ball GD, Bellin ME, Ax RL, First NL (1982). Glycosaminoglycans in bovine cumulus-oocyte complexes:morphology and chemistry. Mol Cell Endocrinol 28:113.
6. Bradley MP, Garbers DL (1983). The stimulation of bovine caudal epididymal sperm forward motility on bovine cumulus-egg complexes in vitro. Biochem Biophys Res Comm 115:777.
7. Lorton SP, First NL (1979). Hyaluronidase does not disperse the cumulus oophorus surrounding bovine ova. Biol Reprod 21:301.
8. Leibfried-Rutledge ML, Critser ES, Eyestone WH, Northey DL, First NL (1987). Developmental potential of bovine oocytes matured in vitro or in vivo. Biol Reprod 36:376.
9. Staigmiller RB, Moor RM (1984). Effect of follicle cells on the maturation and development competence of ovine oocytes matured outside the follicle. Gamete Res 9:221.
10. Lu KH, Gordon I, Gallagher M, McGovern H (1987). Pregnancy established in cattle by transfer of embryos derived from in vitro fertilisation of oocytes matured in vitro. Vet Rec 121:259.
11. Leibfried-Rutledge ML, Critser ES, Parrish JJ, First NL (1989). In vitro maturation and fertilization of bovine oocytes. Theriogenology 31:61-74.

12. Sirard MA, Leibfried-Rutledge ML, Parrish JJ, Ware CM, First NL (1988). The culture of bovine oocytes to obtain developmentally competent embryos. Biol Reprod 39:546.
13. Choudary JB, Gier HT, Marion GB (1968). Cyclic changes in bovine vesicular. J Anim Sci 27:468.
14. Eyestone WH (1989). Factors affecting early bovine development in vitro and in vivo. Ph.D. Thesis, University of Wisconsin-Madison.
15. Boland MP (1984). Use of rabbit oviduct as a screening tool for the viability of mammalian eggs. Theriogenology 21:126.
16. Eyestone, WH, Leibfried-Rutledge ML, Northey DL, Gilligan BG, First NL (1987). Culture of one- and two-cell bovine embryos to the blastocyst stage in the ovine oviduct. Theriogenology 28:1.
17. Rexroad CE Jr, Powell AM (1988). Co-culture of ovine ova with oviductal cells in medium 199. J Anim Sci 66:947.
18. Eyestone WH, First NL (1989). Co-culture of early bovine embryos with oviductal tissue or in conditioned medium. J Reprod Fertil 85:715.
19. Barnes FL (1988). Characterization of the onset of embryonic control and early development in the bovine embryo. Ph.D. Thesis, University of Wisconsin-Madison.
20. Eyestone WH, First NL (1988). Cell cycle analysis of early bovine embryos. Theriogenology 29:243.
21. Hamada S, Fujita S (1983). DAPI-staining improved for quantitative cytofluorometry. Histochemistry 29:219.
22. Crosby IM, Gandolfi F, Moor RM (1988). Control of protein synthesis during early cleavage of sheep embryos. J Reprod Fertil 82:769.
23. Bolton VN, Oades PJ, Johnson MH (1984). The relationship between cleavage, DNA replication and gene expression in the mouse 2-cell embryo. J Embryol exp Morph 79:139.
24. Frei RE, Schultz GA, Church RB (1989). Qualitative and quantitative changes in protein synthesis occur at the 8-16-cell stage of embryogenesis in the cow. J Reprod Fertil 86:637.

THE EXPRESSION OF GROWTH FACTOR LIGANDS AND RECEPTORS IN PREIMPLANTATION MOUSE EMBRYOS[1]

Daniel A. Rappolee, Karin S. Sturm, Gilbert A. Schultz,*
Roger A. Pedersen, and Zena Werb

Laboratory of Radiobiology and Environmental Health
and the Department of Anatomy, University of California
San Francisco, California 94143
and the
*Department of Medical Biochemistry
University of Calgary, Calgary, Alberta T2N 4N1

ABSTRACT Control of growth and differentiation during mammalian embryogenesis may be regulated by growth factors from embryonic or maternal sources. Using a novel method for RNA phenotype analysis based on production of cDNA, followed by enzymatic amplification of specific fragments with the polymerase chain reaction, we have examined simultaneous expression of growth factor ligand and receptor transcripts in preimplantation mouse embryos. In mouse blastocysts, five growth factor genes and three growth factor receptors were expressed, but six growth factor genes were not expressed. Thus, the preimplantation mouse embryo may produce and respond to endogenous growth factors as well as to maternal growth factors.

INTRODUCTION

Because mouse preimplantation embryos grow and differentiate in the absence of exogenous factors, endogenous factors must sustain the embryo during the first six cleavage divisions (1). These early cleavage

---

[1]This work was supported by the Office of Health and Environmental Research, U.S. Department of Energy, contract no. DE-AC03-76-SF01012, by the National Institutes of Health National Research Service Award 5 T32 ES07106 from the National Institute of Environmental Health Sciences, by NIH grants HD 23539 and HD 23651, by Medical Research Council (Canada) grant MT 4854, and by the Alberta Heritage Foundation for Medical Research.

divisions serve two unique functions in mammals: the generation of progenitors of the extraembryonic membranes and the generation of the embryonic anlagen that will form the inner cell mass and hence the embryo proper. Fate maps indicate that after implantation mammalian gastrulation and neurulation may be mechanistically and morphologically similar to that of nonmammalian vertebrates, such as *Xenopus*. However, unlike the abbreviated *Xenopus* cleavage cell cycles that precede gastrulation, mouse preimplantation embryos have near-normal cell cycle times (2), which may be regulated by growth factors. The paradigm for intercellular regulation of growth and differentiation is the interaction of growth factor ligands and receptors.

A body of indirect evidence indicates that preimplantation embryos make growth factors. First, cultured preimplantation embryos produce transforming growth factor-like bioactivity that promotes anchorage-independent growth (3). It is not known if this activity is due to transforming growth factor (TGF)-α, TGF-β, platelet-derived growth factor (PDGF), or some unknown anchorage-independent growth-stimulating activity. Second, shortly after implantation in the uterus, mouse embryos produce basic fibroblast growth factor (bFGF) protein (4,5), TGF-α protein and mRNA (6,7), TGF-β protein (8), insulin-like growth factor (IGF)-II polypeptide (9), and *int-2* mRNA (10), and human embryos produce IGF-II transcripts (11). These factors have been implicated only in the later phases of postimplantation growth and differentiation, and their presence may not indicate growth factor production by the preimplantation embryo. Other evidence for growth factor production in early mammalian embryogenesis comes from teratocarcinoma cells, which are thought to be similar to the primitive ectoderm (12). The differentiated progeny of some teratocarcinoma lines are also equivalent to endodermal cells of the blastocyst (13). Undifferentiated teratocarcinoma cells produce PDGF protein (14) and three protein stem cell growth factors (15). Differentiated teratocarcinoma cells also respond to nerve growth factor (NGF) (16), IGF-II (17), epidermal growth factor (EGF) (18), and PDGF (19). Whether these transformed cells accurately reflect the conditions in preimplantation embryos is not known, because transformation may be caused by improper expression of growth factors or receptors in these lines.

Direct evidence for growth factor transcripts in low copy number in preimplantation embryos has been heretofore impossible to obtain. Localization of mRNA transcripts in embryos by *in situ* hybridization is difficult (20), and no data for growth factor transcripts have been published using preimplantation embryos. Thousands of embryos are required to detect high-copy-number transcripts, such as histone or actin, by RNA blotting analysis (21,22). We have used a novel method for assaying, unambiguously and simultaneously, the accumulation of several growth factor transcripts in small numbers of mouse embryos (23,24).

RESULTS

We developed a method for phenotyping mRNA in small numbers (1-100) of mouse embryos (Fig. 1). The method consists of three linked techniques: a microadaptation of the guanidine thiocyanate/CsCl technique for isolating whole RNA (24), followed by reverse transcription using oligo(dT) or specific antisense oligonucleotide priming, and cDNA amplification by polymerase chain reaction. The products of the first-strand cDNA synthesis are divided and amplified separately by sequence-specific primers (Fig. 2) to produce a phenotype of growth factor transcripts. The method is highly sensitive; messages from a single cell, a single embryo, or as few as 10 synthetic RNA transcripts can be detected (24). The primers bracket a target sequence of diagnostic length of 0.2-0.5 kb and are chosen for (a) sequence specificity, (b) potential diagnostic

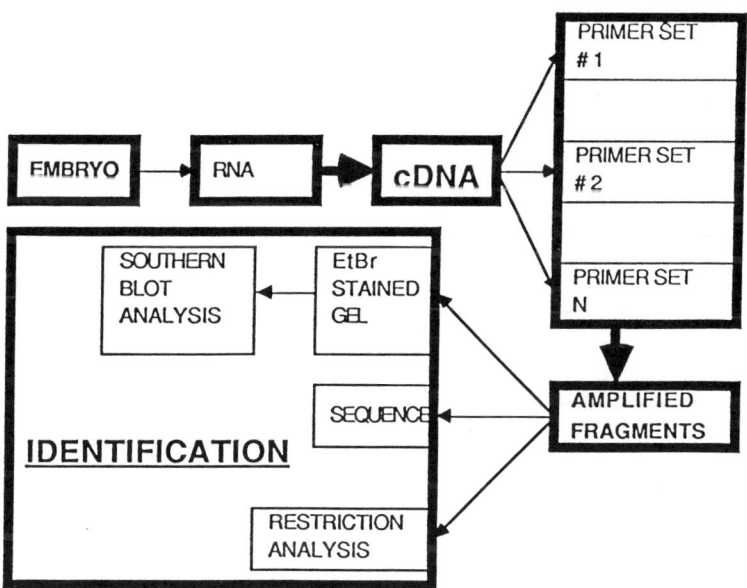

FIGURE 1. Schematic representation of the steps involved in detecting the simultaneous mRNA phenotype in single embryos.

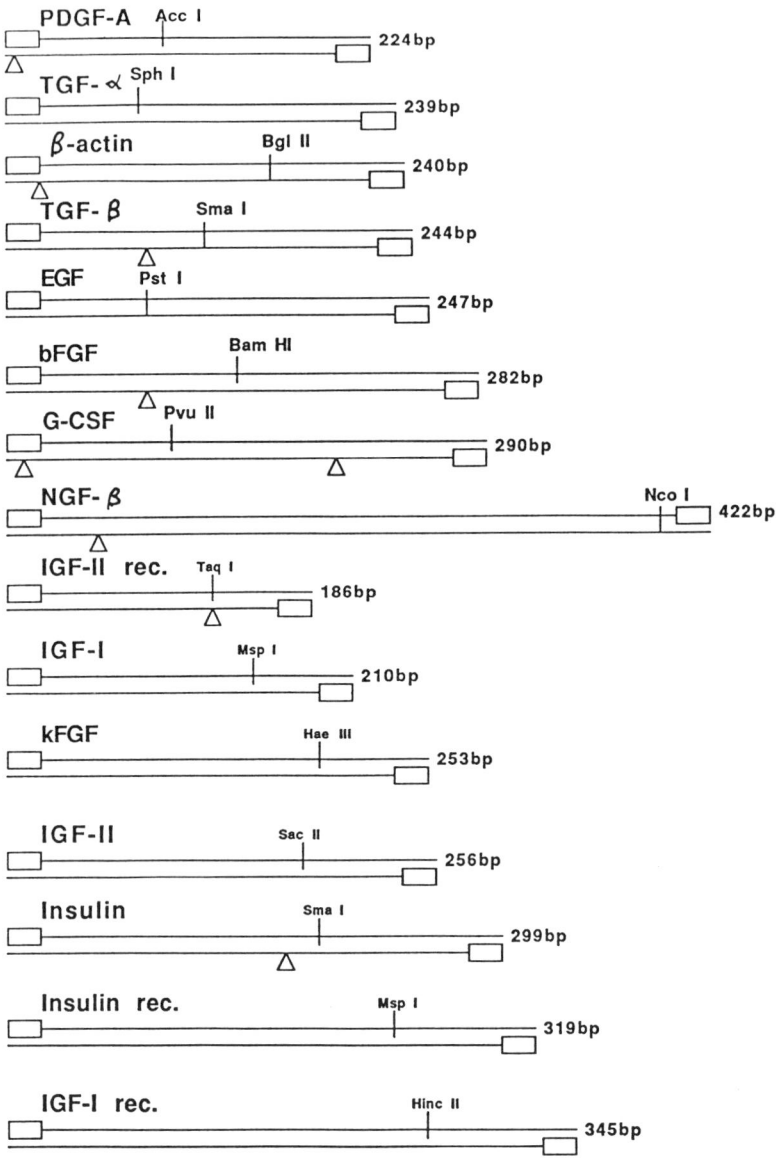

FIGURE 2. Primer and PCR fragment design. Boxes indicate 20-26mer sequence-specific primers, vertical slashes indicate restriction sites, and triangles indicate intron positions.

traits, such as restriction endonuclease sites or cDNA inclusion, and (c) unique interaction with only the mature, processed transcript (Fig. 2).

We found that preimplantation mouse embryos synthesize TGF-α, TGF-β1, PDGF-A, Kaposi's sarcoma-type fibroblast growth factor (kFGF), and IGF-II transcripts but do not synthesize insulin, IGF-I, NGF-β, granulocyte-colony-stimulating factor (G-CSF), bFGF, or EGF transcripts (Table 1). These transcripts fall into three temporal classes: (a) not transcribed at any time before implantation; (b) present as maternal transcripts, destroyed, and resynthesized as zygotic transcripts (TGF-α, PDGF-A); or (c) transcribed only as zygotic transcripts (TGF-β1, IGF-II). A fourth class of transcripts represented by β-actin and the metalloproteinase stromelysin seem to be present throughout preimplantation development. Transcripts for IGF-I receptor, IGF-II receptor, and insulin receptor are present after the activation of the zygotic genome (unpublished observation; Table 2).

TABLE 1
GROWTH FACTOR LIGAND mRNA TRANSCRIPTS FOUND IN PREIMPLANTATION MOUSE EMBRYOS BY RT-PCR (E = unfertilized egg, 2,4,8 = cell number in embryo, B = blastocyst, EC = undifferentiated Nulli cells, and nd = not determined)

Growth Factor	E	2	4	8	B	EC
TGF-α	+	-	+	+	+	+
IGF-II	-	+	+	+	+	+
kFGF	nd	nd	+	+	+	+
PDGF-A	+	-	+	+	+	+
TGF-β1	-	+	+	+	+	+
EGF	nd	nd	nd	nd	-	-
IGF-I	-	-	-	-	-	+
Insulin	-	-	-	-	-	-
bFGF	nd	nd	nd	nd	-	-
NGF-β	nd	nd	nd	nd	-	+
G-CSF	nd	nd	nd	nd	-	-

TABLE 2
GROWTH FACTOR RECEPTOR mRNA TRANSCRIPTS FOUND IN PREIMPLANTATION MOUSE EMBRYOS BY RT-PCR (E = unfertilized egg, 2,4,8 = cell number in embryo, B = blastocyst, and EC = Nulli cells)

Growth Factor Receptor	E	2	4	8	B	EC
IGF-I-R	-	-	-	+	+	+
IGF-II-R	-	+	+	+	+	+
Insulin-R	-	-	-	+	+	+

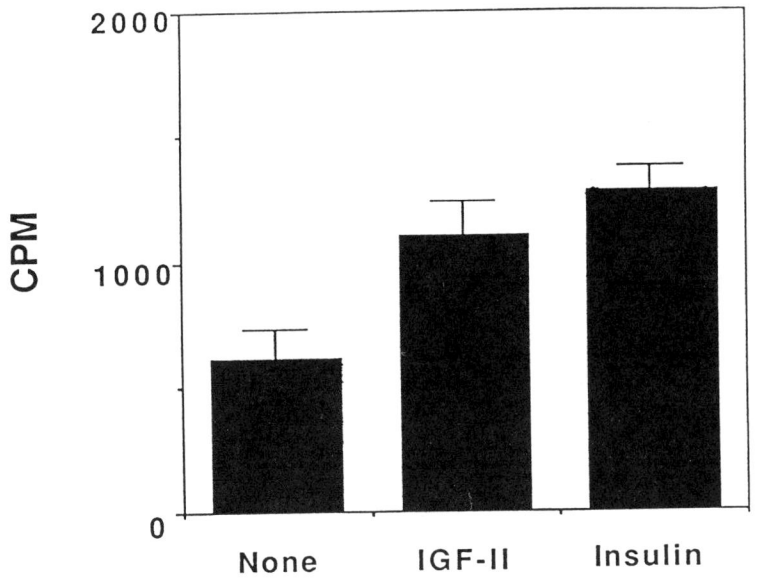

FIGURE 3. Changes in incorporation of [^{35}S]methionine into cellular proteins of mouse blastocysts cultured with growth factors.

Because not all growth factor transcripts are translated, we next asked whether a set of growth factor genes were expressed as polypeptides in mouse blastocysts. We found immunochemical evidence for the translation products of the three ligands TGF-α, TGF-β1, and PDGF (23).

Because IGF-I receptor, IGF-II receptor, and insulin receptor were transcribed (Table 2), but of the corresponding ligands only IGF-II ligand was transcribed, we tested for function at the three receptors. The growth factor-mediated stimulation of incorporation of radiolabeled amino acids into protein in cultured blastocysts was measured by modifying the technique of Harvey and Kaye (25). IGF-II, insulin (Fig. 3), and IGF-I (data not shown) stimulated significant increases in protein synthesis in cultured mouse blastocysts.

DISCUSSION

The relatively autonomous behavior of mammalian embryos before implantation suggests that their growth and differentiation may be sustained by endogenous growth factors. Using a novel technique for RNA phenotyping, we have provided direct evidence that TGF-α, TGF-β, PDGF-A, kFGF, and IGF-II genes are expressed in mouse blastocysts. We found that PDGF-A transcripts are present in unfertilized, ovulated oocytes, disappear by the eight-cell stage, and reappear as zygotic transcripts at the early blastocyst stage. TGF-α was also seen from unfertilized oocyte to blastocyst. Although growth factor transcription can take place without translation (26), we detected TGF-α, TGF-β1, and PDGF antigens in the blastocysts by immunocytochemistry. We also found that blastocysts accumulate IGF-I receptor, IGF-II receptor, and insulin receptor mRNA, and that these receptors may be functional.

SINGLE-CELL mRNA PHENOTYPING BY REVERSE TRANSCRIPTION-POLYMERASE CHAIN REACTION

We have developed a new method for analyzing the mRNA phenotype of preimplantation mouse embryos. The combination of reverse transcription followed by polymerase chain reaction allows detection of < 100 transcripts of c and transcripts from a single cell (24,27). In addition, the products of the reverse transcription can be divided and primed separately in the PCR, a process we call reverse transcription-polymerase chain reaction (RT-PCR). This method has been used to detect growth factor transcripts in small numbers of preimplantation embryos in which polypeptide growth factor transcripts have previously been undetected. Thus, this method provides a powerful new approach for determining the mRNA phenotype of rare tissues and cells at early stages of embryonic development, before the onset of terminal differentiation, and when only the most abundant transcription products can be detected by other, less sensitive methods.

Production of amplified fragments in the PCR is approximately described by a derivation of the equation for bacterial growth, $N = N_0 E^n$, where E is the efficiency of the reaction (which has a range of 1 to 2, corresponding to a doubling efficiency of 0-100%; for 50% efficiency, $E = 1.5$), n is PCR cycle number, N_0 is number of input cDNA molecules, and N is final number of amplified fragments. Using this equation and the results from a single-cell experiment (23,24,27), we estimate that each cDNA molecule can produce 10^9 fragments and that the average efficiency of PCR (over 60 cycles) for cDNA is 35-55%. From this equation it is evident that small changes in PCR efficiency produce large changes in N. We are able to resolve three-fold differences in input RNA and, therefore, threefold differences in RNA between internally controlled samples. Reverse transcription followed by PCR can therefore provide a means of rapid qualitative transcriptional phenotyping of low cell or transcript numbers. The quantification of each transcript requires an internal standard curve generated from cRNA that has a sequence identical to the endogenous mRNA over the range from the 5´ primer to the poly-A tail. The technique also allows rapid cross-species cloning and sequencing and production of high specific activity DNA probes (23,28).

EMBRYONIC GROWTH FACTORS AND RECEPTORS

The mouse embryo grows autonomously for only the first six divisions, at which time it interacts with, and implants in, the wall of the uterus. The controlling influences on these first divisions have not been studied. Zygotic gene transcription begins after the first cell division. To date, there has been no direct proof that preimplantation mouse embryos synthesize growth factors, although there are indirect results suggesting that embryos can bind and express specific growth factors. For example, preimplantation mouse embryos specifically bind EGF (13), and peri-implantation mouse embyros cultured for two days produce transforming growth factor-like biological activity (3). We have found that TGF-α, TGF-β1, kFGF, PDGF-A, and IGF-II genes are expressed in mouse blastocysts and that this transcription is selective: blastocysts do not transcribe genes for EGF, bFGF, NGF-β, G-CSF, insulin, or IGF-I. It is likely that the TGF-α, PDGF, and TGF-β described here, individually or in combination, account for the transforming growth factor-like activity described by Rizzino (3). We have also found that preimplantation mouse embryos transcribe insulin receptor, IGF-I receptor, and IGF-II receptor mRNA, and that these receptors may be functional in these embryos.

There is ample evidence for growth factor production and/or responsiveness in postimplantation fetal development. First, postimplantation rodent embryos at 7.5 days or later stages of gestation produce and/or respond to TGF-β (8), NGF (29), TGF-α (6,7), bFGF (4,5), IGF-II (9,17), and *int-2*, which has a sequence related to bFGF (10). Although most growth factors are not expressed in mouse embryos until the organogenesis phase (day 9), *int-2* (10) and TGF-α (7) are expressed by 7.5 days, only two days after implantation. Second, embryonal carcinoma stem cell lines, which resemble the pluripotential cells of the inner cell mass in the preimplantation embryo, and their immediate progeny of endodermal lineage produce NGF-β (16,30), PDGF (19), bFGF (14), and TGF-α (31). However, transformation of these stem cell lines may involve irregular expression of growth factor receptors or ligands. We found no NGF-β expression in blastocysts, whereas a trace of NGF-β has been found in F9 and PCC4 embryonal carcinoma cell lines (30). Third, embryonal carcinoma stem cells and/or their progeny specifically bind EGF (18), IGF-II (31), PDGF (19), and NGF (16). Taken together, these findings suggest a role for growth factors in early embryonic growth development in mammals.

Growth factors have been implicated in the embryonic development of diverse nonmammalian species. bFGF and TGF-β appear to be morphogens for inducing mesoderm at the blastulation stage in *Xenopus* (32-35); TGF-β-like and EGF-like molecules may influence *Drosophila* development (36-38); EGF-like molecules may influence nematode development (39); and IGF-II-like and FGF-like molecules may influence chick development (40,41). These growth factors can induce differentiation, as in *Xenopus* (33-35), or induce both differentiation and mitosis, as in chick (40,42). In the frog, growth factors may influence differentiation before the eighth cell division (34). In the mouse, maternal growth factor transcripts are replaced by zygotic growth factor transcripts before the sixth cell division. Early development in mouse has several other properties that distinguish it from that of frog. First, the egg is small, has little yolk, and quickly activates its zygotic transcription after fertilization (43). Second, the mouse has 10- to 12-hour cell cycle times (2,44) after the first two cell cycles. These cycles have the normal G1/S/G2/M periods, in contrast to the early cell divisions of frog, which lack G1 and G2 (45). The presence of G1 and G2 in cleavage-stage mouse embryos may allow transcription of growth factors, as well as the opportunity to be influenced by growth factors.

The accumulation patterns of growth factor transcripts in

preimplantation mouse embryos fall into two classes. In one class, including PDGF-A and TGF-α in mouse, maternal transcripts apparently disappear and are resynthesized in the zygote; frog FGF (34) and PDGF-A transcripts act similarly (46). In the second class, transcripts survive the breakdown of maternal mRNA that is initiated during meiotic maturation, becomes quite dramatic in the 2-cell embryo, and continues up to the blastocyst stage (43). TGF-β is an example of the second class. Similarly, in *Xenopus*, the TGF-β-like Vg-1, which is localized to the vegetal hemisphere and may play a role in mesoderm induction, persists throughout early development (33). A comparable physiological role for TGF-β1 or TGF-β2 in mouse is not known.

What does the presence of these three early growth factor transcripts imply about their function in mouse embryos? We can separate growth factor functions by two criteria: direction and action. The direction can be within the embryo or between the embryo and the mother. The action can be to influence mitosis and/or differentiation. An intraembryonic premitogenic function is suggested by our findings that IGF-II ligand and its receptors, the IGF-I and IGF-II receptors, are expressed and that the receptors are functional (47). An intraembryonic mitogenic function is suggested by the coincidental production, by the autonomous blastocyst, of the three growth factors belonging to a factor subset that sustains anchorage-independent growth (48). The onset of growth factor transcription from the zygotic genome in mouse roughly coincides with, or precedes, the differentiation of totipotent inner cell mass cells into primitive ectoderm and endoderm. Several lines of evidence indicate that embryonic factors are directed at maternal tissue. The strongest evidence is the prolongation of corpus luteum lifespan in sheep by an embryonic protein thought to be ovine trophoblast protein-1 (49). This protein, which binds endometrial receptors and is the major translation product of ovine trophoblasts, was recently cloned and found to be highly homologous to a secreted polypeptide factor, α_{11}-interferon (50). Our evidence for embryonic-maternal communication is more circumstantial. First, TGF-α and TGF-β are known to be angiogenic (51,52), and the highest density of uterine capillary beds is opposite the implanting blastocyst (53). In addition, the uterine environment is hypoxic (54), a condition that promotes wound healing cells to produce angiogenic factors (55). Finally, at the time of implantation there is a surge of estrogen that increases EGF receptor expression in uterus several-fold (56); TGF-α is an EGF receptor-binding ligand. Taken together, these data indicate that embryonic growth factors may contribute to the induction of early angiogenesis and decidualization of the uterus. A possible maternal-to-embryo influence is represented by the functional IGF-I and insulin receptor transcripts and proteins and the lack of IGF-I and insulin ligand transcripts and endogenous polypeptides.

We speculate that IGF-II produced by mouse preimplantation embryos may modulate embryonic growth through IGF-I or IGF-II receptors, which are functional in early mouse embryos. The functional insulin and IGF-I receptors in mouse preimplantation embryos may transduce maternally derived signals because the embryo does not transcribe insulin or IGF-I mRNA.

REFERENCES

1. Biggers JD (1971). New observations on the nutrition of the mammalian oocyte and the preimplantation embryo. In Blandau RJ (ed): "Biology of the Blastocyst," Chicago: University of Chicago Press, p 319.
2. Pedersen RA (1986). Potency, lineage, and allocation in preimplantation mouse embryos. In Rossant J, Pedersen RA (eds): "Experimental Approaches to Mammalian Embryonic Development," Cambridge: Cambridge University Press, p 3.
3. Rizzino A (1985). Early mouse embryos produce and release factors with transforming growth factor activity. In Vitro Cell Dev Biol 21:531.
4. Risau W (1986). Developing brain produces an angiogenesis factor. Proc Natl Acad Sci USA 83:3855.
5. Risau W, Ekblom P (1986). Production of a heparin-binding angiogenesis factor by the embryonic kidney. J Cell Biol 103:1101.
6. Lee DC, Rochford R, Todaro GJ, Villarreal LP (1985). Developmental expression of rat transforming growth factor-α mRNA. Mol Cell Biol 5:3644.
7. Twardzik DR (1985). Differential expression of transforming growth factor-α during prenatal development of the mouse. Cancer Res 45:5413.
8. Heine U, Munoz EF, Flanders KC, Ellingsworth LR, Lam HY, Thompson NL, Roberts AB, Sporn MB (1987). Role of transforming growth factor-β in the development of the mouse embryo. J Cell Biol 105:2861.
9. D'Ercole AJ, Underwood LE (1980). Ontogeny of somatomedin during development in the mouse. Dev Biol 79:33.
10. Jakobovits A, Shackleford GM, Varmus HE, Martin GR (1986). Two proto-oncogenes implicated in mammary carcinogenesis, *int-1* and

int-2, are independently regulated during mouse development. Proc Natl Acad Sci USA 83:7806.
11. Scott J, Cowell J, Robertson ME, Priestley LM, Wadey R, Hopkins B, Pritchard J, Bell GI, Rall LB, Graham CF, Knott TJ (1985). Insulin-like growth factor-II gene expression in Wilms' tumour and embryonic tissues. Nature 317:260.
12. Martin GR, Evans MJ (1975). Differentiation of clonal lines of teratocarcinoma cells: Formation of embryoid bodies *in vitro*. Proc Natl Acad Sci USA 72:1441.
13. Adamson ED (1986). Cell-lineage-specific gene expression in development. In Rossant J, Pedersen RA (eds): "Experimental Approaches to Mammalian Embryonic Development," Cambridge: Cambridge University Press, p 321.
14. van Veggel JH, van Oostwaard TMJ, de Laat SW, van Zoelen EJJ (1987). PC13 embryonal carcinoma cells produce a heparin-binding growth factor. Exp Cell Res 169:280.
15. Jakobovits A, Banda MJ, Martin GR (1985). Embryonal carcinoma-derived growth factors: Specific growth-promoting and differentiation-inhibiting activities. In Feramisco J, Ozanne B, Stiles C (eds): "Growth Factors and Transformation," Cold Spring Harbor, N.Y.: Cold Spring Harbor Laboratory, p. 393.
16. Liesi P, Rechardt L, Wartiovaara J (1983). Nerve growth factor induces adrenergic neuronal differentiation in F9 teratocarcinoma cells. Nature 306:265.
17. Heath JK, Rees AR (1985). Growth factors in mammalian embryogenesis. In "Growth Factors in Biology and Medicine," Ciba Foundation Symp. 116:3.
18. Adamson ED, Hogan BLM (1984). Expression of EGF receptor and transferrin by F9 and PC13 teratocarcinoma cells. Differentiation 27:152.
19. Rizzino A, Bowen-Pope D (1985). Production of PDGF-like factors by embryonal carcinoma cells and response to PDGF by endoderm-like cells. Dev Biol 110:15.
20. Han VKM, Hunter ES III, Pratt RM, Zendegui JG, Lee DC (1987). Expression of rat transforming growth factor alpha mRNA during development occurs predominantly in maternal decidua. Mol Cell Biol 7:2335.
21. Piko L, Clegg KB (1982). Quantitative changes in total RNA, total poly(A), and ribosomes in early mouse embryos. Dev Biol 89:362.

22. Giebelhaus DH, Heikkila JJ, Schultz GA (1983). Changes in the quantity of histone and actin messenger RNA during the development of preimplantation mouse embryos. Dev Biol 98:148.
23. Rappolee DA, Brenner CA, Schultz R, Mark D, Werb Z (1988). Developmental expression of PDGF, TGF-α, and TGF-β genes in preimplantation mouse embryos. Science 241:1823.
24. Rappolee DA, Wang A, Mark D, Werb Z (1989). Novel method for studying mRNA phenotypes in single or small numbers of cells. J Cell Biochem 39:1.
25. Harvey MB, Kaye PL (1988). Insulin stimulates protein synthesis in compacted mouse embryos. Endocrinology 122:1182.
26. Assoian RK, Fleurdelys BE, Stevenson HC, Miller PJ, Madtes DK, Raines EW, Ross R, Sporn MB (1987). Expression and secretion of type β transforming growth factor by activated human macrophages. Proc Natl Acad Sci USA 84:6020.
27. Rappolee DA, Mark D, Banda MJ, Werb Z (1988). Wound macrophages express TGF-α and other growth factors in vivo: Analysis by mRNA phenotyping. Science 241:708.
28. Brenner CA, Adler RR, Rappolee DA, Pedersen RA, Werb Z (1989). Genes for extracellular matrix-degrading metalloproteinases and their inhibitor, TIMP, are expressed during early mammalian development. Genes Dev, in press.
29. Levi-Montalcini R, Booker B (1960). Destruction of the sympathetic ganglia in mammals by an antiserum to a nerve-growth protein. Proc Natl Acad Sci USA 46:384.
30. Dicou E, Houlgatte R, Brachet P (1986). Synthesis and secretion of β-nerve growth factor by mouse teratocarcinoma cell lines. Exp Cell Res 167:287.
31. Heath JK, Shi W-K (1986). Developmentally regulated expression of insulin-like growth factors by differentiated murine teratocarcinomas and extraembryonic mesoderm. J Embryol Exp Morph 95:193.
32. Slack JMW, Darlington BG, Heath JK, Godsave SF (1987). Mesoderm induction in early Xenopus embryos by heparin-binding growth factors. Nature 326:197.
33. Weeks DL, Melton DA (1987). A maternal mRNA localized to the vegetal hemisphere in Xenopus eggs codes for a growth factor related to TGF-β. Cell 51:861.
34. Kimelman D, Kirschner M (1987). Synergistic induction of mesoderm by FGF and TGF-β and the identification of an mRNA coding for FGF in the early Xenopus embryo. Cell 51:869.

35. Rosa F, Roberts AB, Danielpour D, Dart LL, Sporn MB, David IB (1988). Mesoderm induction in amphibians: The role of TGF-β2-like factors. Science 239:783.
36. Padgett RW, St. Johnston RD, Gelbart WM (1987). A transcript from a Drosophila pattern gene predicts a protein homologous to the transforming growth factor-β family. Nature 325:81.
37. Wharton KA, Johansen KM, Xu T, Artavanis-Tsakonas S (1985). Nucleotide sequence from the neurogenic locus notch implies a gene product that shares homology with proteins containing EGF-like repeats. Cell 43:567.
38. Hafen E, Basler K, Edstroem J-E, Rubin GM (1987). Sevenless, a cell-specific homeotic gene of Drosophila, encodes a putative transmembrane receptor with a tyrosine kinase domain. Science 236:55.
39. Greenwald I (1985). *lin-12*, a nematode homeotic gene, is homologous to a set of mammalian proteins that includes epidermal growth factor. Cell 43:583.
40. Bell KM (1986). The preliminary characterization of mitogens secreted by embryonic chick wing bud tissues *in vitro*. J Embryol Exp Morph 93:257.
41. Goldin GV, Opperman LA (1980). Induction of supernumerary tracheal buds and the stimulation of DNA synthesis in the embryonic chick lung and trachea by epidermal growth factor. J Embryol Exp Morph 60:235.
42. Engstrom W, Bell KM, Clemmons DR (1987). Expression of the insulin like growth factor II gene in the developing chick limb. Cell Biol Int Rep 11:415.
43. Schultz GA (1986). Utilization of genetic information in the preimplantation mouse embryo. In Rossant J, Pedersen, RA (eds): "Experimental Approaches to Mammalian Embryonic Development," Cambridge: Cambridge University Press, p 239.
44. Molls M, Zamboglou N, Streffer C (1983). A comparison of the cell kinetics of pre-implantation mouse embryos from two different mouse strains. Cell Tissue Kinet 16:277.
45. Graham CF, Morgan RW (1966). Changes in the cell cycle during early amphibian development. Dev Biol 14:439.
46. Mercola M, Melton DA, Stiles CD (1988). Platelet-derived growth factor A chain is maternally encoded in Xenopus embryos. Science 241:1223.

47. Rappolee DA, Schultz GA, Pedersen RA, Sturm K, Werb Z (1989). An endogenous growth factor-receptor circuit in preimplantation mammalian development? J Cell Biochem Suppl 13B:200.
48. Anzano M, Roberts AB, Sporn MB (1986). Anchorage independent growth of primary rat embryo cells is induced by platelet derived growth factor and inhibited by type-beta transforming growth factor. J Cell Physiol 126:312.
49. Weitlauf HM (1988). Biology of implantation. In Knobil E, Neill J (eds): "Physiology of Reproduction," New York: Raven Press, p 231.
50. Imakawa K, Anthony RV, Kazemi M, Marotti KR, Polites HG, Roberts RM (1987). Interferon-like sequence of ovine trophoblast protein secreted by embryonic trophectoderm. Nature 330:377.
51. Schreiber AB, Winkler ME, Derynck R. (1986). Transforming growth factor-α. A more potent angiogenic mediator than epidermal growth factor. Science 232:1250.
52. Roberts AB, Sporn MB, Assoian RK, Smith JM, Roche NS, Wakefield LM, Heine UI, Liotta LA, Falanga V, Kehrl JH, Fauci AS (1986). Transforming growth factor type β: Rapid induction of fibrosis and angiogenesis *in vivo* and stimulation of collagen formation *in vivo*. Proc Natl Acad Sci USA 83:4167.
53. Williams MF (1948). The vascular architecture of the rat uterus is influenced by estrogen and progesterone. Am J Anat 83:274.
54. Yochim JM (1981). Intrauterine oxygen tension and metabolism of the endometrium during the preimplantation period. In Blandau RJ (ed): "The Biology of the Blastocyst," Chicago: University of Chicago Press, p 363.
55. Knighton DR, Hunt TK, Scheuenstuhl H, Halliday BJ, Werb Z, Banda MJ (1983). Oxygen tension regulates the expression of angiogenesis factor by macrophages. Science 221:1283.
56. Mukku VR, Stancel GM (1985). Regulation of epidermal growth factor receptor by estrogen. J Biol Chem 260:9820.

CHANGES IN RNA AND PROTEIN SYNTHESIS DURING DEVELOPMENT OF THE PREIMPLANTATION MOUSE EMBRYO.[1]

Gilbert Schultz, Wendy Dean, Ann Hahnel, Nancy Telford Daniel Rappolee*, Zena Werb*, and Roger Pedersen*

        Department of Medical Biochemistry
        University of Calgary
        Calgary, Alberta T2N 4N1 Canada and

  * Laboratory of Radiobiology and Environmental
    Health and Department of Anatomy
    University of California, San Francisco
    San Francisco, CA 94143

ABSTRACT The transition from maternal to embryonic control in the mouse embryo occurs at the 2-cell stage. Major changes in the pattern of protein synthesis occur at this stage as maternal mRNAs decay and new mRNAs derived from transcription of the zygote genome become translated. The small ribonucleoprotein particles involved in pre-mRNA splicing during transcriptional activation of the zygote genome appear to be maternal in origin. They are released from the germinal vesicle into the oocyte cytoplasm during meiotic maturation but relocalize to nuclei during the first cleavage division following fertilization. Many genes begin to be expressed at the 2-cell and later stages of preimplantation development. We have used the mRNA phenotyping technique to demonstrate that alkaline phosphatase mRNAs are first detectable at the 2-cell stage and that transcripts encoding insulin receptors are not detectable until the 8-cell stage. These results document that the cellular machinery for mRNA processing is present and functional in early mouse embryos and reveal the presence of rare transcripts previously undetectable by less sensitive methods.

---

1. This work was supported by grant MT4854 from MRC (Canada), the Alberta Heritage Foundation for Medical Research, NIH (HD 23539 and HD 23651), by contract DE-AC03-76-SF01012 from the US DOE-OHER and by a National Research Service Award (5T32 ES07106) from the NIEHS.

INTRODUCTION

The developmental program begins during the process of oogenesis. In the mouse, all classes of RNA are synthesized in the growing oocyte and several changes in the pattern of protein synthesis occur as the oocyte matures [1,2]. At the time of ovulation, the mouse oocyte is arrested at second meiotic metaphase and is transcriptionally inactive. The first cleavage is controlled almost exclusively by macromolecules accumulated in the egg [3].

During the first cleavage, significant qualitative changes in the pattern of radiolabeled proteins occur because some maternal mRNA molecules become translationally activated, some polypeptides become post-translationally modified and newly synthesized mRNAs from the zygote genome begin to be translated [3-6]. A switch from maternal to zygote genome control of mouse embryo development occurs during this first cell cycle and is accompanied by changes in polyadenylation of maternal mRNAs [7-9], loss of about 70% of the poly A(+) RNA [10] and major reductions in the abundance of a number of specific mRNAs [7-14]. By the 2-cell stage, the synthesis of all major classes of RNA has been re-initiated and RNA content increases progressively as new transcription continues and cell number increases up to the blastocyst stage [3,10,15].

The acquisition of the capacity for transcription and post-transcriptional processing in the 2-cell mouse embryo has been demonstrated by plasmid microinjection experiments [16] and by the fact that paternally derived gene products are detectable by the late 2-cell stage [17]. These observations imply that the early embryo must contain the machinery for processing pre-mRNA. The cellular machinery that mediates mRNA processing is composed of snRNA molecules that are found in ribonucleoprotein complexes (snRNPs) in the cell nucleus [18]. In this paper we demonstrate by in situ hybridization and immunofluorescence microscopy that there is a substantial amount of snRNA in oocytes but, in contrast to maternal mRNA, it is not degraded in the one-cell to two-cell transition period. Rather, snRNAs and snRNPs are released into the cytoplasm of the unfertilized egg during germinal vesicle breakdown and meiotic maturation and subsequently relocalize to pronuclei of the fertilized egg and nuclei of the first two blastomeres in the zygote [19,20]. Thus, the pre-mRNA that is synthesized during transcriptional activation of the zygote genome may be processed by snRNA of maternal origin.

Transcriptional activation of the embryonic genome at the 2-cell stage is a major regulatory event that results in the production (in addition to maternal-type gene transcripts) of many new mRNA species that are not represented in detectable amounts in egg RNA [21]. Temporally regulated patterns of gene activation from the 2-cell stage onward ultimately lead to the differentiation of the first two phenotypically distinct cell types, trophectoderm and inner cell mass, at the blastocyst stage. This process includes synthesis of transcripts encoding particular enzymes or growth factors and receptors that can influence cell proliferation and differentiation [22]. Such transcripts are often short-lived and of low copy number within cells. They can, however, be detected within RNA isolated from just one or a few cells or preimplantation mouse embryos through a novel method of mRNA phenotype analysis. This method is based upon production of cDNA by reverse transcription (RT) followed by the polymerase chain reaction (PCR) to amplify specific target sequences [23,24]. We have used this method to examine the expression of two genes, alkaline phosphatase (ALP) and insulin receptor (Ins-R), that are each expressed in a temporally specific manner following activation of the zygote genome. Transcripts for ALP were first detected at the 2-cell stage whereas transcripts for Ins-R were first detected at the 8-cell stage.

METHODS

Animals and Embryo Recovery

Random-bred CD1 mice (Charles River Breeding Laboratories) were used. Procedures for superovulation, collection of ovarian oocytes with intact germinal vesicles, recovery of ovulated oocytes and recovery of preimplantation stage embryos have been described previously [12,13,19]. Injection of human chorionic gonadotropin (HCG) was used to induce ovulation. Ovulated oocytes were collected at 14-16 h post-HCG, pronuculear zygotes at 20-22 h post-HCG, two-cell embryos at 44-46 h post-HCG, eight-cell embryos at 68-70 h post-HCG, morulae at 80-83 h post-HCG and early blastocysts (about 32 cells) at 92 to 96 h post-HCG.

Recombinant DNA Probes

Histone. A 900 bp segment of mouse genomic DNA derived by Sal I and Eco Rl cleavage and containing the coding region for amino acids 58 to 135 and adjacent flanking region of a histone H3.2 gene was ligated into pBR322 [25].

U2 snRNA. A 300 bp genomic fragment containing a 188 nucleotide mouse U2 gene [26] was cloned into pGEM-1. Linearization of the recombinant plasmid with Eco Rl produced an antisense RNA upon transcription with Sp6 polymerase and the sense-strand (control) RNA following linearization with Hind III and transcription with T7 polymerase [20].

ALP. A full length (2.4 kb) clone encoding a liver/bone/kidney type ALP was isolated from a cDNA library constructed from poly A(+) RNA from nullipotent embryonal carcinoma SCC1 cells and subcloned into a Bluescript (Stratagene) vector [27].

Ins-R. A full length 4.1 kb human insulin receptor cDNA was subcloned into the Eco Rl site of pUC12 [28] and was provided as a gift by Dr. A. Ullrich (Genentech).

RNA Extraction and Hybridization Procedures

Total RNA was extracted from oocytes and embryos in the presence of 10 µg of tRNA carrier as described previously [12,13,19]. Procedures for resolution of RNAs in formaldehyde-agarose gels and transfer to Nytran (Schleicher and Schull) membranes, hybridization and washes were identical to those reported previously [19].

The histone H3 probe was radiolabeled by nick-translation in the presence of α-^{32}[P]-dCTP (3000 Ci/mmol, NEN Dupont). ALP and Ins-R cDNAs were radiolabeled in the presence of α-[^{32}P]-dCTP by the random-priming method. Radiolabeled RNA probes for Northern blots or in situ hybridization with U2 snRNA clones were generated from transcription vectors in the presence of α-^{35}[S]-UTP (1300 Ci/mmol, NEN Dupont) as described previously [19].

For RT-PCR, RNA was extracted from oocytes or embryos by the procedure described above except that 10 µg of E. coli ribosomal RNA (Boehringer-Mannheim) was used as carrier.

In Situ Hybridization.

Procedures used for fixation, embedding, hybridization, post-hybridization washes and autoradiography were exactly as described previously [19].

Immunofluorescence Microscopy.

An international reference human anti-Sm serum [29] that recognizes U1, U2, U4, U5 and U6 snRNPs was diluted 1:10 prior to experimental use. All other procedures for fixation and permeabilization of mouse oocytes and embryos, application of primary and secondary antibodies, observation of stained material and photography were conducted as described previously [19]. Localization of nuclei was achieved through staining with the DNA-specific fluorescent dye diamidinophenylindole (DAPI) as described previously [30].

mRNA Analysis by Reverse Transcription (RT) and the Polymerase Chain Reaction (PCR).

The procedures for RT of oocyte and embryo RNA and amplification of oligo-dT primed cDNAs with specific primer pairs by PCR were exactly as described in previous studies [22-24].
Primer pairs were synthesized in the Regional DNA Synthesis Laboratory, Department of Medical Biochemistry, The University of Calgary, Calgary, Alberta, Canada. For ALP, the 5' oligonucleotide sequence was complementary to nucleotides 1283-1302 of the mouse ALP cDNA [27] and had the sequence: (5') GTGGATACACCCCCCGGGGC (3'). The 3' oligonucleotide was located at nucleotides 1591-1612 and had the sequence: (5') GGTCAAGGTTGGCCCCAATGCA (3'). The predicted target sequence bracketed by the primer pairs was 330 nucleotides in length. For insulin receptor, a target sequence of 319 nucleotides was bracketed between the 5' oligonucleotide primer [(5') ACTGACCTCATGCGCATGTGCTGG(3')] located between nucleotide 3808-3831 of the human Ins-R cDNA sequence [28] and the 3' oligonucleotide primer [(5')GCCCGTTTTTCTTGCCTCCGTTCAT(3')] sequence located between nucleotides 4102-4126 of the Ins-R cDNA clone.
Products of the PCR reaction (6 μl) were resolved electrophoretically on agarose gels composed of 3% NuSieve

GTG and 1% Seakem ME (FMC Bioproducts), stained with ethidium bromide and photographed [22-24]. For restriction enzyme analysis, amplified PCR products were ethanol precipitated and resuspended in appropriate restriction enzyme buffer. Digested and undigested products were separated on 6% acrylamide gels and stained with ethidium bromide. For Southern blots, PCR fragments were resolved on 2% agarose or 6% acrylamide gels and electrophoretically transferred to Nytran membranes for hybridization.

Protein Synthetic Studies

Oocytes and embryos were incubated in medium containing L-^{35}[S]-methionine (1 mCi/ml, 1200 Ci/mmol, Amersham) for 2 hours. Washed individual oocytes or embryos were lysed in dissociation buffer and radiolabeled polypeptides were resolved by one-dimensional sodium dodecyl-sulfate polyacrylamide gel electrophoresis (SDS-PAGE) and fluorography as described previously [31].

RESULTS AND DISCUSSION

The ovulated mouse egg contains a functional translation apparatus due to accumulation of rRNA, mRNA, tRNA and ribosomes during oogenesis [1,2]. During the first 24 hours of postfertilization development, there is little change in net protein synthetic rate, protein turnover or total protein content [32,33]. The qualitative pattern of protein synthesis, however, changes markedly during the early cleavage period [Fig. 1]. Many of the changes in protein synthetic pattern at the 2-cell stage are due to post-translational modification, differential mRNA activation or differential polypeptide turnover of maternal gene products. All of these mechanisms are involved in the well-described changes that occur within a protein complex of about 35 kD [Fig. 1] during the first 24 hours after fertilization [4-6]. Other changes at the 2-cell stage are dependent upon new transcription because they do not appear in fertilized eggs cultured to the 2-cell stage in the presence of α-amanitin. The 70 kD complex of heat shock proteins [hsp 70, Fig. 1] is represented within this transcriptionally-dependent set of polypeptides [4,34].

Quantitatively, protein synthesis remains at a relatively low rate from fertilization to the 8-cell stage but there is a progressive increase in synthetic rate in subsequent stages accompanying the transition of the morula to the blastocyst [32]. Qualitative changes in protein synthetic profile between the 8-cell and blastocyst stages are not as marked as those which occur during the transition from maternal to zygote control at the 2-cell stage [Fig. 1] but a number of stage-specific and inner cell mass (ICM) and trophectoderm-specific polypeptides have also been demonstrated in two-dimensional fluorographs of newly synthesized proteins from morulae and blastocysts [35-37].

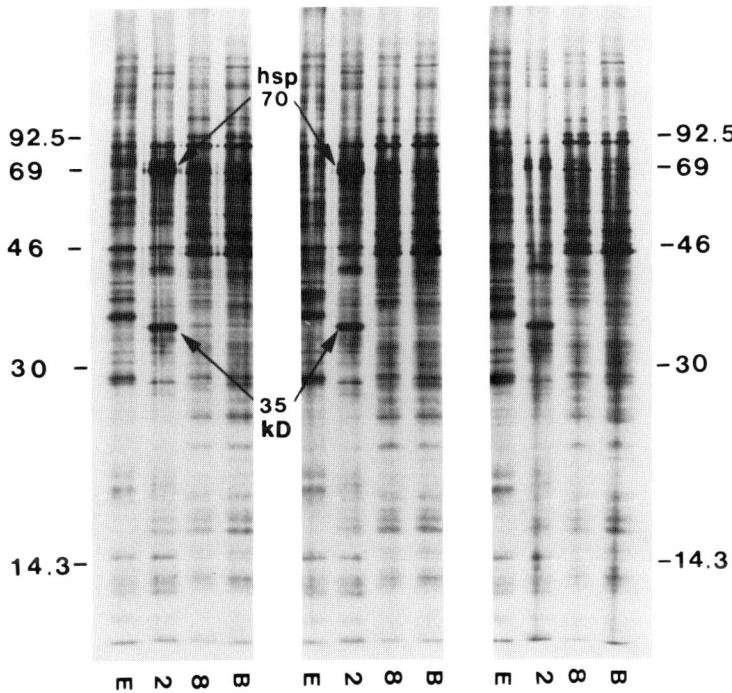

FIGURE 1 Stage-dependent synthesis of proteins during early mouse development. Unfertilized oocytes (E) and 2-cell (2), 8-cell (8) or blastocyst stage (B) embryos were incubated in medium containing L-[^{35}S]-methionine at 37°C for 2 h. About 2500 cpm of acid-precipitable radioactivity from three sets of individual oocytes or embryos was analyzed by SDS-PAGE electrophoresis and fluorography. The position of [^{14}C]-methylated standards is shown in each margin with numerical values in kD. Arrows identify the position of 35 kD and 68 - 70 kD hsp proteins.

The switch from maternal to zygote genome control during the first cleavage is accompanied by the degradation of maternal mRNA. This is demonstrated in one example by the marked reduction in histone H3 mRNA in 2-cell embryos as compared to unfertilized oocytes [Fig. 2]. Similar reductions in mRNA abundance have been reported for several other transcripts [7-14]. By the 2-cell stage, transcription of mRNA from the embryonic genome has been activated and histone mRNAs re-accumulate along with other mRNAs during the general increase in RNA synthetic activity and RNA content that occurs between the 2-cell and blastocyst stage [Fig. 2, ref. 3,10].

The capacity to generate translatable mRNAs from the embryonic genome requires that functional snRNAs and snRNPs for processing of pre-mRNA are also present at the 2-cell stage. It has been recently shown that there are substantial amounts of U1, U2, U4, U5 and U6 snRNA in oocytes [19,20] but, in contrast to the bulk of maternal mRNA, they are not degraded during the 1- to 2-cell transition. The relative abundance of U2 RNA is shown in Fig. 2. The general profile of the other snRNAs involved in pre-mRNA processing is similar over this developmental interval [19,20]. There are about 1×10^6 copies of each of the snRNAs in oocytes and 2-cell embryos. Following transcriptional activation at the 2-cell stage, the snRNA content increases to about 1×10^7 molecules per blastocyst [20].

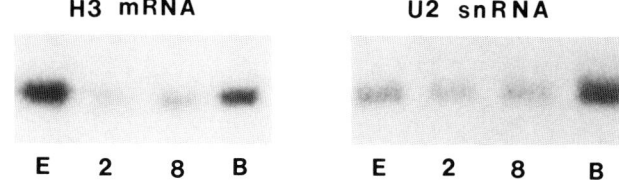

FIGURE 2. Northern blot analysis of RNA from early mouse embryos. RNA was extracted from unfertilized oocytes (E), 2-cell (2), 8-cell (8) or blastocyst (B) stage embryos, resolved on denaturing agarose gels and transferred to membranes for hybridization. For histone H3 analysis, RNA was extracted from about 1000 eggs or embryos at each stage of development and hybridized with a probe radiolabeled with [^{32}P] by nick-translation (sp. act. of 2×10^8 cpm/µg). Autoradiographic exposure time was 2 days. For U2 snRNA analysis, RNA was extracted from about 200 eggs or embryos and the blot was hybridized with a [^{35}S]-labeled antisense RNA probe (1.4×10^8 cpm/µg) and exposure time was 12 days.

Localization of snRNAs and snRNPs has been examined by in situ hybridization and immunofluorescence microscopy. The distribution of U2 snRNA is shown in the left panel of Fig. 3 but the pattern is also the same for other snRNAs [19,20]. The distribution of the protein component (Sm antigen) of U1, U2, U4, U5 and U6 snRNPs has been examined with anti-Sm serum [right panel, Fig. 3]. Both methods demonstrate that snRNAs and snRNPs are primarily localized to the germinal vesicle in the preovulatory oocyte but are released and diluted within the cytoplasm of the oocyte following germinal vesicle breakdown and meiotic maturation. They subsequently relocalize to both pronuclei following fertilization and both nuclei of the two blastomeres generated by the first cleavage division. Since the amount of snRNA is constant during the first cleavage division, the small amount of pre-mRNA that is first synthesized during transcriptional activation of the zygote genome is probably spliced and processed by snRNPs of maternal origin. In later stages of preimplantation development the snRNAs and snRNPs remain predominantly localized to interphase nuclei [Fig. 3].

The transition from maternal to embryonic control results in major qualitative and quantitative changes in the mRNA populations in embryos from the 2-cell stage onward. Some of the mRNAs, such as those encoding histones, are present in both the maternal mRNA set and the embryonic mRNA set, albeit at different concentrations [Fig. 2]. However, in an analysis of clones from a small cDNA library prepared from 2-cell embryos, a number of the cloned sequences that were expressed at the 2-cell stage were not represented within egg mRNA [21]. That is, a second set of mRNAs begin to be expressed at the 2-cell stage that are embryo-specific. In addition, there are yet other mRNAs derived from temporally regulated genes that are not expressed until later stages of preimplantation development [3].

Northern and slot blot techniques have been useful in the study of the expression of abundant mRNA transcripts during preimplantation development including, for example, those for actin [12], histone [13], zona pellucida [14], intracisternal A particles [21], cytochrome oxidase I and II [21] and tissue plasminogen activator [9]. In brief, all of these mRNA species are present in the range of 10 to 1000 fg/oocyte or embryo (1 X 10^4 to about 5 X 10^5 molecules/oocyte or embryo) and each represents about 0.01 to 0.5% of the total mRNA pool. If several hundred oocytes

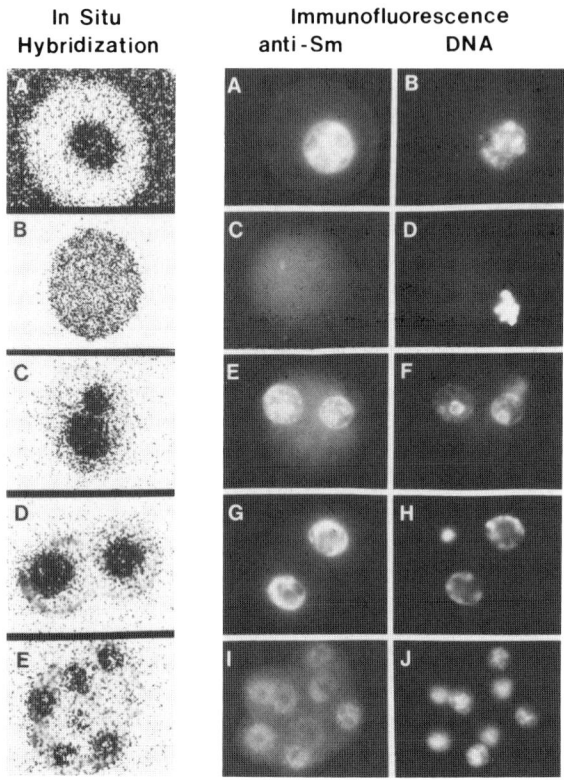

FIGURE 3. Localization of snRNA and snRNPs in mouse oocytes and preimplantation embryos. Left Panels: In situ hybridization. (A) Preovulatory oocyte with intact germinal vesicle; (B) Unfertilized oocyte; (C) Fertilized egg with two pronuclei; (D) 2-cell embryo; (E) 8-cell embryo. Oocytes or embryos were fixed in Carnoy's, embedded in paraffin, sectioned at 5 um thickness, hybridized in situ with a [^{35}S]-labeled antisense U2 snRNA probe, dipped in emulsion and subjected to autoradiography for 6 days. Right Panels: Anti-Sm snRNP immunofluorescent localization. Anti-Sm staining is shown in the left column of photomicrographs and DNA fluorescent localization (DAPI staining) is shown in the right column. (A,B) Preovulatory oocyte with germinal vesicle; (C,D) Unfertilized oocyte arrested at second meiotic metaphase; (E,F) Pronuclear stage; (G,H) 2-cell stage. (I,J) 8-cell stage.

or embryos are used for each RNA preparation, pg quantities of the particular mRNA species are generated and are sufficient for detection in blot-based hybridization protocols. Nuclease protection experiments, by virtue of the fact that the probe is provided in excess in a solution hybridization reaction, are at least 10-fold more sensitive than a Northern blot experiment and are useful in detection of variant expression within small multigene families [13,19]. Despite this, these approaches are still limited by the amount of biological material that is practically available in this system and are not sufficiently sensitive to detect low copy number mRNAs that may be involved in growth control and differentiation in the early mouse embryo.

Recently, a modification of the PCR reaction incorporating an RT step has been used to demonstrate the presence of mRNA transcripts for various growth factors and receptors in a small number of macrophages [23] and preimplantation stage mouse embryos [24]. This mRNA phenotyping method can be used to detect fewer than 100 mRNA molecules in as little as 10 pg of total RNA. We applied this method to analyze the expression of alkaline phosphatase and insulin receptor genes because studies at the protein level suggested they are temporally regulated during early mouse development; however, we were unable to detect transcripts by hybridization of appropriate cDNA clones to Northern blots of oocyte and embryo RNA. The sensitivity of the RT-PCR method was established by making serial dilutions of total cytoplasmic RNA from nullipotent embryonal carcinoma (EC) SCC1 cells in amounts ranging from 100 ng to 1 pg. The RNAs were subjected to two cycles of reverse transcription with oligo-dT priming and then amplified with primers specific for ALP or Ins-R cDNA sequences. In each case, the amplified fragment was detectable by ethidium bromide staining after 60 cycles of PCR even in samples that contained as little as 10 pg of total RNA, the approximate amount present in a single EC cell [Fig. 4A, 4B].

There are four genes within the human genome that code for ALPs, a family of homodimeric, glycosylated, membrane-bound enzymes that catalyze the hydrolysis of phosphate esters at alkaline pH. The four major types correspond to the isozymes found in germ-cell, placenta, intestine and liver/bone/kidney respectively [38]. The germ-cell, placental and intestinal forms are more than 90% identical in amino acid sequence [38] but the liver/bone/kidney type is quite different (only 52% identity with the placental form at the amino acid level) and maps to a different chromosome [39]. The genes contain 11 exons interrupted by 10 introns and code for polypeptides that are about 510 amino acids long through mRNAs that are about 2.4 kb in length [38-42].

While high levels of ALP activity are present in mouse bone, liver, intestine, kidney and placenta [43], ALP activity is also prominent in early embryonic cells, primordial germ cells, teratocarcinoma cells and embryonal carcinoma cells [44-47]. In the early embryo, ALP enzymatic activity is first detectable at the late-two to four-cell stage and progressively increases through the 8-cell and blastocyst stages [48-50].

On the basis of enzymatic studies and immunological data, the kind of ALP in mouse teratocarcinoma and embryonal carcinoma cells appears to be similar to the liver/bone/kidney type and is distinguishable from the intestinal form [51]. We have recently isolated and sequenced a full-length 2.4 kb ALP cDNA from a nullipotent EC cell library [27] and it contains only minor differences from another mouse ALP clone isolated earlier from a placental cDNA library [52]. Both mRNAs code for a protein that is 90% identical to the human liver/bone/kidney enzyme but only 55% identical to the human placental and intestinal enzymes [27,38].

We have used PCR methodology to demonstrate that the mRNA for the EC cell ALP isozyme is present in preimplantation mouse embryos. The target sequence bracketed by the oligonucleotide primers and that should be amplified by RT-PCR is expected to be 330 nucleotides in length. Such a PCR fragment is, indeed, detectable from the 2-cell stage onward and the amount increases progressively to the blastocyst stage [Fig. 4C]. Identity of the amplified PCR fragment was further verified by the fact that it was cleaved by Hha I into two fragments, 193 and 137 base pairs in length, as predicted from the cDNA sequence [27] and by the fact that it could be hybridized with the EC cell ALP cDNA clone when electro-blotted onto Nytran membrane [Fig. 4E].

The oligonucleotide primers were designed to bridge the last two introns (0.7 and 0.5 kb in length) that separate coding regions in the liver/bone/kidney ALP gene [41], thus distinguishing properly spliced mRNA from genomic DNA or unprocessed pre-mRNA. Since the ALP fragment amplified by PCR is the correct length for spliced mRNA with contiguous

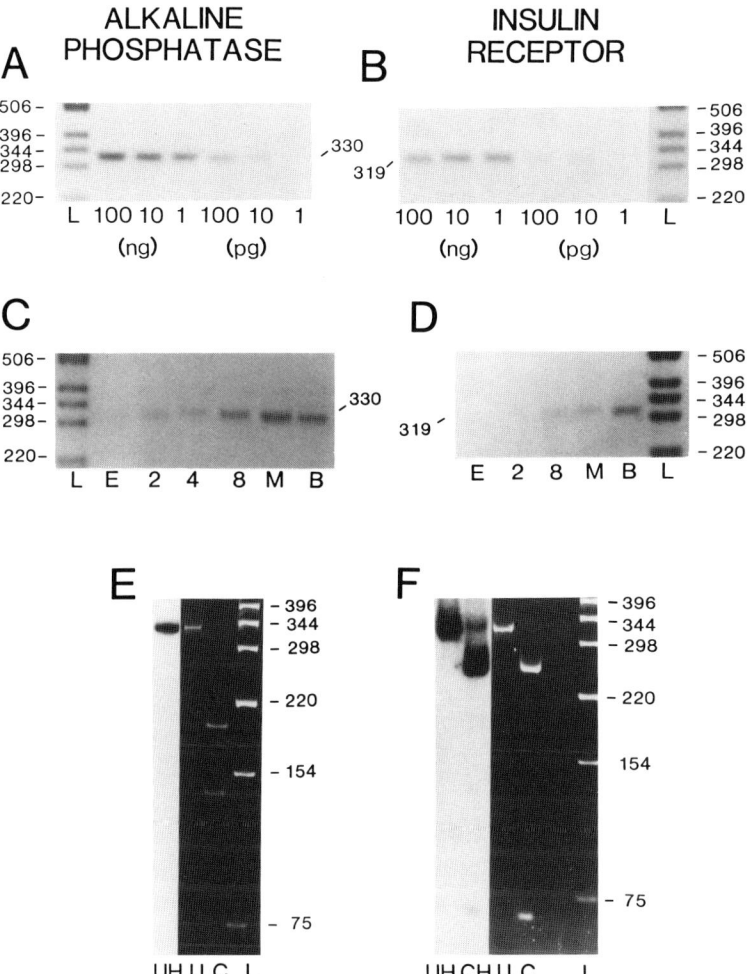

FIGURE 4. Expression of alkaline phosphatase and insulin receptor transcripts in early mouse embryos. Panel A and B; sensitivity of the RT-PCR method for detection of ALP mRNA (Panel A) or Ins-R mRNA (Panel B). Total RNA from nullipotent EC cells was serially diluted from 100 ng to 1 pg, reverse-transcribed by oligo-dT priming, amplified by 60 cycles of PCR with primers specific for ALP or Ins-R cDNA and the products resolved by agarose gel electrophoresis. L indicates markers (in base pairs) from a small molecular weight DNA ladder. Panel C; ALP mRNA in preimplantation embryos. RNA from 110 ovulated oocytes (E), 225 two-cell

Continued on page 40.

exon coding sequences, this result also confirms that the mRNA splicing machinery is functional at the 2-cell stage and thereafter.

Other ALP isozymes also appear to be expressed during early mouse development. For example, it has been shown that antibodies directed toward human placental ALP react with mouse blastocysts but not F9 EC cells [50,53]. It will be possible to search for transcripts from other mouse ALP genes when additional sequence information becomes available.

There is much current research on the role of growth factors and receptors in cell proliferation, differentiation and morphogenesis during early mouse development. Expression of the insulin receptor gene is of particular interest. Insulin is bound in a developmentally regulated fashion to the cells of preimplantation mouse embryos from the late 8-cell to morula stage onward [54]. Autoradiographic studies have confirmed that insulin-binding is receptor mediated [55]. Moreover, addition of exogenous insulin to culture medium at physiological levels leads to

---

Continued from page 39.
embryos (2), 128 four-cell embryos (4), 112 eight-cell embryos (8), 60 eight-to sixteen-cell morulae (M) and 106 blastocysts (B) were reverse-transcribed and one-tenth of the RT product was treated as in Panel A. The 330 bp amplified PCR fragment is first detectable at the 2-cell stage. Panel D; Ins-R mRNA in preimplantation embryos. RNA from 111 oocytes (E), 82 two-cell embryos (2), 118 8-cell embryos (8), 66 eight to 16-cell morulae (M), and 72 blastocysts (B), was reverse transcribed and one-tenth of the reverse transcription product was treated as in Panel B. The 319 bp amplified PCR fragment is first detectable at the 8-cell stage. Panel E; Verification of the identity of the ALP PCR fragment. The position of uncut 330 bp fragment (U) and the 193 and 137 bp fragments derived from Hha I cleavage (C) are shown in relation to markers from a small molecular weight DNA ladder (L). In the left lane, uncut PCR fragment was electro-blotted to Nytran membrane and hybridized with a radiolabeled ALP probe (UH). Panel F; Verification of identity of the Ins-R PCR fragment. Uncut (U, 319 bp) and Hinc II cleavage fragments (C, 253 and 66 bp) are shown in relation to the markers in the DNA ladder (L). In the left lanes, both uncut (UH) and cut (CH) fragments were transferred to membranes and hybridized with a labeled human Ins-R cDNA probe.

the stimulation of both glucose transport [56] and protein synthesis [57] in compacted morulae and blastocysts but not in early cleavage stages of development.

We have designed oligonucleotide primers for RT-PCR based on the human Ins-R cDNA sequence [28] to search for Ins-R transcripts in early mouse embryos. The expected product, 319 base pairs in length, was first detected in RNA from the 8-cell stage and increased in quantity up to the blastocyst [Fig. 4D]. The amplified PCR product contained the diagnostic Hinc II restriction endonuclease site that, following cleavage, yielded the expected fragments of 253 and 66 base pairs [Fig. 4F]. In addition, when blotted, the fragments hybridized with the radiolabeled human cDNA probe [Fig. 4F]. These results confirm the earlier reports that insulin receptors are first expressed at the 8-cell stage of mouse development.

The type of analyses reported above for ALP and Ins-R gene expressions along with those of several other genes [22], will eventually lead to a detailed molecular phenotypic map for each stage of preimplantation mouse development. It should then be possible to use antisense techniques [58,59] or other genetic manipulations to bring about a block in the expression of a temporally regulated gene at the appropriate point in the developmental program. The developmental consequence of aberrant expression could then be studied in embryos maintained in culture or re-transferred to the reproductive tract of appropriate recipients. Ultimately, it should be possible to assess the function of these and other genes expressed in early mouse embryos on subsequent developmental processes.

REFERENCES

1. Bachvarova R (1985). Gene expression during oogenesis and oocyte development in mammals. In Browder LW(ed): "Developmental Biology. A Comprehensive synthesis, Vol. 1, Oogenesis." New York: Plenum Press, p 453.
2. Schultz RM (1986). Molecular aspects of mammalian oocyte growth and maturation. In Rossant J, Pedersen RA (eds): "Experimental Approaches to Mammalian Embryonic Development." Cambridge: Cambridge University Press, p. 195.
3. Schultz GA (1986). Utilization of genetic information in the preimplantation mouse embryo. In Rossant J, Pedersen RA (eds): "Experimental Approaches to

Mammalian Embryonic Development." Cambridge: Cambridge University Press, p 239.
4. Flach G, Johnson MH, Braude PR, Taylor RAS, Bolton VN (1982). The transition from maternal to embryonic control in the 2-cell mouse embryo. EMBO Journal 1: 681.
5. Bolton VN, Oades PJ, Johnson MH (1984). The relationship between cleavage, DNA replication and gene expression in the mouse 2-cell embryo. J Embryol exp Morph 79: 139.
6. Howlett SK, Bolton VN (1985). Sequence and regulation of morphological and molecular events during the first cell cycle of mouse embryogenesis. J Embryol exp Morph 87: 175.
7. Clegg KB, Piko L (1983). Poly(A) length, cytoplasmic adenylation and synthesis of poly(A)+ RNA in early mouse embryos. Dev Biol 95: 331.
8. Paynton BV, Rempel R, Bachvarova R (1988). Changes in state of adenylation and time course of degradation of maternal mRNAs during oocyte maturation and early embryonic development in the mouse. Dev Biol 129: 304.
9. Huarte J, Belin D, Vassali A, Strickland S, Vassali JD (1987). Meiotic maturation of mouse oocytes triggers the translation and polyadenylation of dormant tissue-type plasminogen activator mRNA. Genes and Dev 1: 1201.
10. Piko L, Clegg KB (1982). Quantitative changes in total RNA, total poly(A) and ribosomes in early mouse embryos. Dev Biol 89: 362.
11. Mutter GL, Grills GS, Wolgemuth DJ (1988). Evidence for the involvement of the protooncogene c-mos in mammalian meiotic maturation and possibly very early embryogenesis. EMBO Journal 7: 683.
12. Giebelhaus DH, Heikkila JJ, Schultz GA (1983). Changes in the quantity of histone and actin messenger RNA during the development of preimplantation mouse embryos. Dev Biol 98: 148.
13. Graves RA, Marzluff WF, Giebelhaus DH, Schultz GA (1985). Quantitative and qualitative changes in histone gene expression during early mouse development. Proc Natl Acad Sci USA 82: 5685.
14. Philpott CC, Ringuette MR, Dean J (1987). Oocyte-specific expression and developmental regulation of ZP-3, the sperm receptor of the mouse zona pellucida. Dev Biol 121: 568.
15. Clegg KB, Piko L (1983). Quantitative aspects of RNA

synthesis and polyadenylation in 1-cell and 2-cell mouse embryos. J Embryol exp Morph 74: 169.
16. Chen HY, Trumbauer ME, Ebert KM, Palmiter RD, Brinster RJ (1986). Developmental changes in the response of mouse eggs to injected genes. In Bogorad L (ed): "Molecular Developmental Biology." New York: Alan R Lis Inc., p 149.
17. Sawicki JA, Magnuson T, Epstein CJ (1982). Evidence for expression of the paternal genome in the two-cell mouse embryo. Nature 294: 450.
18. Maniatis T, Reed R (1987). The role of small nuclear ribonucleoprotein particles in pre-mRNA splicing. Nature 325: 673.
19. Lobo SM, Marzluff WF, Seufert AC, Dean WL, Schultz GA, Simerly C, Schatten G (1988). Localization and expression of U1 RNA in early mouse development. Dev Biol 127: 349.
20. Dean WL, Seufert AC, Schultz GA, Prather RS, Simerly C, Schatten G, Pilch DR, Marzluff WF (1989). The small nuclear RNAs for pre-mRNA splicing are coordinately regulated during oocyte maturation and early embryogenesis in the mouse. Manuscript submitted for publication.
21. Taylor KD, Piko L (1987). Patterns of mRNA prevalence and expression of B1 and B2 transcripts in early mouse embryos. Development 101: 877.
25. Werb Z, Schultz GA, Pedersen R, Sturm K and Rappolee DA (1989). Growth factor and growth factor receptor gene expression in peri-implantation mouse embryos. J Cell Biochem Suppl. 13B: 192.
23. Rappolee DA, Mark D, Banda MJ, Werb Z (1988). Wound macrophages express TGF-α and other growth factors in vivo: analysis by mRNA phenotyping. Science 241: 708.
24. Rappolee DA, Brenner CA, Schultz R, Mark D, Werb Z (1988). Developmental expression of PDGF, TGF-α, and TGF-β genes in preimplantation mouse embryos. Science 241: 1823.
25. Sittman DB, Chiu I, Pan C, Cohn RH, Kedes LH, Marzluff WF (1981). Isolation of two clusters of mouse histone genes. Proc Natl Acad Sci USA 78: 4078.
26. Nojima H, Kornberg RD (1983). Genes and pseudogenes for mouse U1 and U2 small nuclear RNAs. J Biol Chem 258: 8151.
27. Hahnel AC, Schultz GA (1989). Cloning and characterization of a cDNA encoding alkaline phosphatase in mouse embryonal carcinoma cells. Clinica Chimica Acta, in press.

28. Ullrich A, Bell J, Chen E, Herrera R, Petruzelli LM, Dull TJ, Gray A, Coussens L, Liao Y, Tsubokawa M, Mason A, Seeburg P, Grunfeld C, Rosen O, Ramchandran J (1985). Human insulin receptor and its relationship to the tyrosine kinase family of oncogenes. Nature 313: 756.
29. Tan EM, Fritzler MJ, McDougall JS, McDuffie FC, Nakamura RM, Reichlin M, Reimer CB, Sharp GC, Schur PH, Wilson MR, Winchester RJ (1982). Reference sera for antinuclear antibodies I. Antibodies to native DNA, Sm, nuclear RNP and SS-B/La. Arthritis Rheum 25: 1003.
30. Ito K, McGhee JD, Schultz GA (1988). Paternal DNA strands segregate to both trophectoderm and inner cell mass of the developing mouse embryo. Genes and Dev 2: 929.
31. Hahnel AC, Gifford DJ, Heikkila JJ, Schultz GA (1986). Expression of the major heat shock protein (hsp 70) family during early mouse embryo development. Terat Carc Mut 6: 493.
32. Brinster RL, Wiebold JL, Brunner S (1976). Protein metabolism in preimplanted mouse ova. Dev Biol 51: 215.
33. Merz EA, Brinster RL, Brunner S and Chen HY (1981). Protein degradation during preimplantation development of the mouse. J Reprod Fert 61: 415.
34. Bensaude O, Babinet C, Morange M, Jacob F (1983). Heat shock proteins, first major products of zygotic gene activity in the mouse embryo. Nature 305: 331.
35. Van Blerkom J, Barton SC, Johnson MH (1976). Molecular differentiation in the preimplantation mouse embryo. Nature 259: 319.
36. Braude PR (1979). Control of protein synthesis during blastocyst formation in the mouse. Dev Biol 68: 440.
37. Howe CC, Gmur R, Solter D (1980). Cytoplasmic and nuclear protein synthesis during in vitro differentiation of murine ICM and embryonal carcinoma cells. Dev Biol 74: 351.
38. Millan JL (1988). Oncodevelopmental expression and structure of alkaline phosphatase genes. Anticancer Research 8: 995.
39. Weiss MJ, Henthorn PS, Lafferty MA, Slaughter C, Raducha M, Harris H (1986). Isolation and characterization of a cDNA encoding a human liver/bone/kidney-type alkaline phosphatase. Proc Natl

Acad Sci USA 83: 7182.
40. Millan JL, Manes T (1988). Seminoma-derived Nagao isozyme is encoded by a germ cell alkaline phosphatase gene. Proc Natl Acad Sci USA 85: 3025.
41. Weiss MJ, Kunal R, Henthorn PS, Lamb B, Kadesch T, Harris H (1988). Structure of the human liver/bone/kidney alkaline phosphatase gene. J Biol Chem 263: 12002.
42. Henthorn PS, Raducha M, Kadesch T, Weiss MJ, Harris H (1988). Sequence and characterization of the human intestinal alkaline phosphatase gene. J Biol Chem 263: 12011.
43. Goldstein DJ, Rogers CE, Harris H (1980). Expression of alkaline phosphatase loci in mammalian tissues. Proc Natl Acad Sci USA 77: 2857.
44. Johnson LV, Calarco PG, Siebert ML (1977). Alkaline phosphatase activity in preimplantation mouse embryos. J Embryol exp Morph 14: 83.
45. Chiquoine AD (1954). The identification, origin and migration of the primordial germ cells in the mouse. Anat Rec 118: 135.
46. Bernstine EG, Hooper ML, Grandchamp S, Ephrussi B (1973). Alkaline phosphatase activity in mouse teratoma. Proc Natl Acad Sci USA 70: 3899.
47. Mintz B, Russel ES (1957). Gene-induced embryological modifications of primordial germ cells in the mouse. J Exp Zool 134: 207.
48. Mulnard J, Huygens R (1978). Ultrastructural localization of non-specific alkaline phosphatase during cleavage and blastocyst formation in the mouse. J Embryol exp Morph 44: 121.
49. Izquierdo L, Lopez T, Marticorena P (1980). Cell membrane regions in preimplantation mouse embryos. J Embryol exp Morph 59: 89.
50. Ziomek CA, Lepire ML (1989). Fluorescent histochemical and immunofluorescent localization of cell surface alkaline phosphatase on mouse preimplantation embryos. Development, in press.
51. Hass PE, Wada HG, Herman MM, Sussman HH (1979). Alkaline phosphatase of mouse teratocarcinoma stem cells: immunochemical and structural evidence for its identity as a somatic gene product. Proc Natl Acad Sci USA 76: 1164.
52. Terao M, Mintz B (1987). Cloning and characterization of a cDNA coding for mouse placental alkaline phosphatase. Proc Natl Acad Sci USA 84: 7051.

53. Lepire ML, Ziomek CA (1989). Preimplantation mouse embryos express a heat-stable alkaline phosphatase. Development, in press.
54. Rosenblum IY, Mattson BM, Heyner S (1986). Stage-specific insulin binding in mouse preimplantation embryos. Dev Biol 116: 261.
55. Mattson BM, Rosenblum IY, Smith RM, Heyner S (1988). Autoradiographic evidence for insulin and insulin-like growth factor binding to early mouse embryos. Diabetes 37: 585.
56. Gardner HG, Kaye PL (1984). Effects of insulin on preimplantation mouse embryos. Proc Aust Soc Reprod Biol 16: 107.
57. Harvey MB, Kaye PL (1988). Insulin stimulates protein synthesis in compacted mouse embryos. Endocrinology 116: 261.
58. Bevilacqua A, Erickson RP, Hieber V (1988). Antisense RNA inhibits endogenous gene expression in mouse preimplantation embryos: Lack of double-stranded RNA "melting activity." Proc Natl Acad Sci USA 85: 831.
59. Strickland S, Huarte J, Belin D, Vassalli A, Rickles RJ, Vassalli J-D (1988). Antisense RNA directed against the 3' noncoding region prevents dormant mRNA activation in mouse oocytes. Science 241: 680.

DETECTION OF EGF RECEPTOR PROTEIN AND ACTIVITIES IN EMBRYO CELLS[1]

Eileen D. Adamson * and Mark Mercola **

La Jolla Cancer Research Foundation *
10901 North Torrey Pines Road
La Jolla, California 92037
Dana Farber Cancer Institute **
Harvard Medical School
44 Binney Street
Boston, Massachusetts 02115

DETECTION OF EGF RECEPTOR PROTEIN AND ACTIVITIES IN EMBRYO CELLS This article will outline some recent advances in growth factor receptor protein studies applied to small numbers of embryo cells. Improvements include iodinated EGF binding, autoradiographic localization and EGF receptor tyrosine kinase activity assays combined with antibodies specific to the receptor. Finally, receptor response can be detected by studying gene products further down the signal pathway, namely c-*fos* protein.

RESULTS

EGF Receptors are Expressed on Trophoblast Giant Cells.

The placenta of most mammalian species examined has been used as a source of EGF receptor protein, but the role of the receptor in fetal or placental development remains obscure (1). In order to gain a better understanding, the cell type and the onset of expression of EGF

---

[1]This work was supported by a grant from the U.S. PHS CA 28427

receptors must be known. Radioiodinated-EGF binding sites have been measured by counting radio-labeled placental membranes or by autoradiography of ligand-labeled placental tissue sections (2). All cell types in the placenta appear to express the receptor. One of the precursor cell-types that gives rise to the placenta was also found to bind ^{125}-I-EGF, that is, the trophoblast giant cells that grow out from mouse day 5 blastocyst embryos cultured in vitro.(2)

Trophoblast Giant Cell EGF-Receptors are Functional.

Although many tissues of the fetus express EGF receptors and can respond to excess EGF by down-regulation of their receptors (3,4), this response does not necessarily mean that the signal reaches the nucleus to generate a change in gene expression. A change in gene expression after EGF stimulation can induce either mitosis or increased differentiated expression depending on the cell type. For example, differentiated teratocarcinoma cells such as the murine PSA5E cell line bind EGF with high affinity, but make no detectable gene responses and continue to grow at a high rate.

Trophoblast giant cells that have grown out from blastocysts over 2 days of culture, appear to be responsive to the growth factors present in fetal bovine serum, as demonstrated by staining with antibodies to c-fos protein (Figure 1). This proto-oncogene is invariably the first detectable gene product after a variety of stimuli are introduced into the cellular environment. Other products such as c-myc protein may be detected subsequently to trace the cascade of cellular responses further. It remains to be seen which of the many growth factors in serum elicit this response in these cells, and whether the effect is transient or more sustained as it might be in these extraembryonic cell types (5).

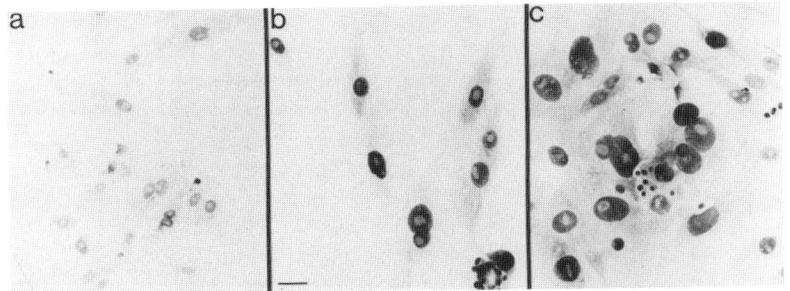

FIGURE 1. Trophoblast cells respond to serum stimulation. Mouse blastocysts were flushed from uteri on the 5th day of gestation and allowed to attach and grow out on gelatinized cover slips in DME with 10% FBS. After 1 day, the serum content was reduced to 0.2% for 24 hrs (A). In B and C, FBS was added to 20% for 45 minutes. All cultures were fixed in methanol for 5 minutes at -20^0 before staining by the avidin-biotin-peroxidase complex procedure using rabbit anti-fos peptide M IgG at $10\mu g/ml$ as the primary antibody. Note that serum stimulates the appearance of fos protein to different degrees in nuclei of several sizes. Bar = 50 microns.

The High Sensitivity of Detection of the Tyrosine Kinase Activity.

The synthesis of EGF receptor protein must start at least as early as the 7th day of gestation judging by the EGF binding activities of the trophoblast giant cell. Egg cylinders on day 7.5 onwards also synthesize a 170 kDa metabolically-labelled polypeptide that immunoprecipitates with polyclonal rabbit antibodies raised to purified mouse liver EGF receptor protein (Figure 2, lanes 9, 10 and 11). In contrast, an antibody that immunoprecipitates the type B PDGF receptor from 3T3 fibroblasts (Figure 2, lane 5) does not produce a comparable band even in day 9.5 embryos. (Figure 2, lane 8). One explanation for this result could be that, at early times in embryogenesis, only type A PDGF receptor is present. This would interact with PDGF-A isoform which is known to be produced by early embryos (6). However, this result could also represent limitations in the technique and we must therefore look for more sensitive methods.

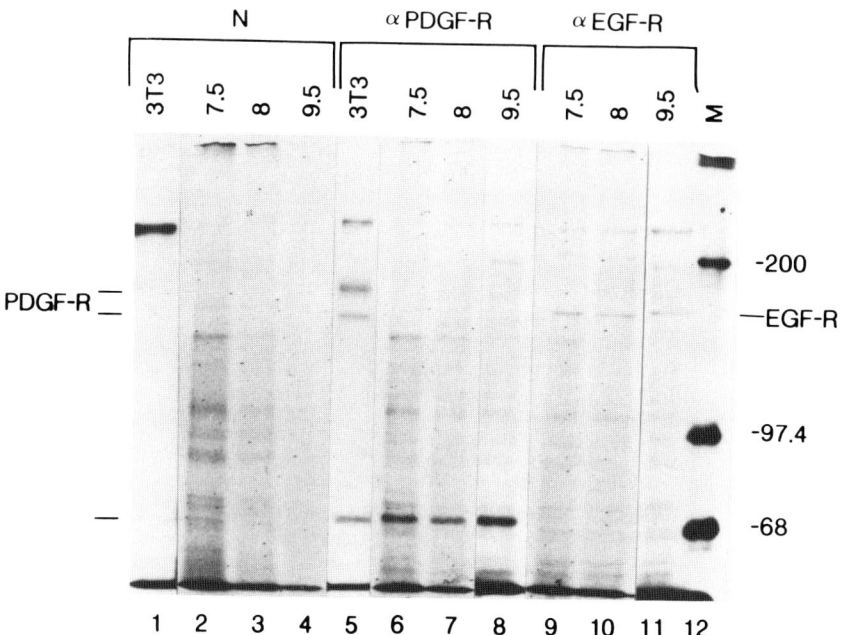

FIGURE 2. Immunoprecipitation of metabolically labelled mouse embryos. Embryos were dissected on day 7.5, 8 and 9.5 of gestation and incubated with ^{35}S-methionine (1mCi/80µl/40 embryos) in methionine - free medium for 4 hours. Lysates in RIPA buffer (1% sodium deoxycholate, 1% Non-idet 40, 0.1% SDS, 50mMTris, 5mM EDTA pH 7.5 were divided into 3 portions for immunoprecipitation with : lanes 1-4, normal rabbit serum; lanes 5-8, antibody to a peptide sequence in mouse type B PDGF receptor; lanes 9-11, anti-EGF receptor. EGF receptor but little or no PDGF receptor protein was detected in embryo lysates. M=markers.

The tyrosine kinase activity of the EGF receptor, discovered by Cohen, et al (7), has great potential in detecting the incorporation of ^{32}P-γ-labelled-ATP into either exogenous substrates such as angiotensin II or into the EGF receptor protein itself by an autocatalytic reaction (8).
A combination of factors raises the sensitivity of detection by this method: the sensitivity of ^{32}P detection

is greater, the enzymatic activity amplifies the signal, and the receptor kinase may incorporate ^{32}P-labelled phosphate into three possible carboxy-terminal tyrosine residues. In addition, the reaction can be performed with crude homogenates, and is very rapid even at 0°C, when other phosphorylation reactions may not be efficient. The specificity of detection is monitored both by the ability of EGF to stimulate the tyrosine kinase activity of the receptor and by analysis of the products after electrophoresis through an SDS-polyacrylamide gel. The EGF receptor is characterized by a band at 170kDa that is stimulated by the addition of EGF in the assay mixture. A further increase in sensitivity can be made by immunoprecipitation of the labelled receptor kinase before gel electrophoresis since this enables the assay to be made on significantly larger volumes of homogenate. This refinement also verifies the nature of the 170 kDa product. We have shown, for instance that EGF receptor protein is readily detectable in 10^6 cells of the OC15 embryonal carcinoma cell-line after EGF stimulation (9). Surprisingly, these cells do not in fact bind exogenous EGF and do not express EGF receptors on their surface. We concluded that these undifferentiated stem cells have intracellular receptors and have speculated that the same may be true for the cells of early embryos.

PERSPECTIVES

These techniques combined with sensitive immunohistochemical procedures such as have been successfully employed by S. Heyner and her colleagues (10,11) to detect insulin receptors, should now allow us to detect, localize and test the activities of EGF receptor protein in very early embryos. The exciting possibility now exists that EGF receptors may function at these early stages. It should also be possible to determine if receptors are intracellular or expressed on the plasma membrane. The other component of any receptor complex is the ligand. What ligands are present in the egg or embryo or in the uterine fluid that would lead to receptor activation and responses? The work of Z. Werb and her colleagues (6) clearly show the presence of transforming growth factor-alpha, an EGF receptor ligand, in the preimplantation embryo. When this data is fully described, we may then have better clues to the functions of the receptor during embryogenesis.

REFERENCES

1. Adamson ED (1979). Developmental activities of the epidermal growth factor receptor. Current Topics in Developmental Biology, (Submitted) Academic Press, N.Y.
2. Adamson ED, Meek J (1984). The ontogeny of epidermal growth factor receptors during mouse development. Dev Biol 103: 62.
3. Adamson ED, Deller MJ, Warshaw JB (1981). Functional EGF receptors on embryonic cells. Nature 291: 656.
4. Adamson ED, Warshaw JB (1981). "Down-regulation" of epidermal growth factor receptors in mouse embryos. Dev Biol 90: 430.
5. Adamson ED, Meek J, Edwards SA (1985). Product of the cellular oncogene, c-fos, observed in mouse and human tissues using an antibody to a synthetic peptide. EMBO J 4: 941.
6. Rappolee DA, Brenner CA, Schultz G, Mark D, Werb Z (1988). Developmental Expression of PDGF, TGF-α and TGF-β genes in preimplantation mouse embryos. Science 241: 1823.
7. Cohen S, Carpenter G, King L (1980). Epidermal Growth Factor - Protein kinase interactions. J Biol Chem 255:4834.
8. Bertics PJ, Gill GN (1985). Self-phosphorylation enhances the protein kinase activity of the epidermal growth factor receptor. J Biol Chem 260: 14642.
9. Weller A, Meek J, Adamson ED (1987). Preparation and properties of monoclonal and polyclonal antibodies to mouse epidermal growth factor (EGF) receptors: Evidence for cryptic EGF receptors in embryonal carcinoma cells. Development 100: 351.
10. Mattson BA, Rosenblum IY, Smith RM, Heyner S (1988). Autoradiographic evidence for insulin and IGF binding to early mouse embryo. Diabetes 37: 585.
11. Heyner S (1989). This volume.

DEVELOPMENTAL REGULATION OF THE KS-FGF ONCOGENE BY EMBRYONAL CARCINOMA CELLS AND EARLY MOUSE EMBRYOS[1]

Angie Rizzino[2], Jay Tiesman, Ronald Hines and David Kelly

Eppley Institute, University of Nebraska Medical Center
Omaha, Nebraska 68105-1065

ABSTRACT We have determined that embryonal carcinoma (EC) cells produce growth factors related to fibroblast growth factor (FGF) and that the levels of this growth factor are reduced after differentiation occurs. To identify the gene that codes for this growth factor, we have performed northern blot analysis with probes for 5 different members of the FGF family of growth factors, including KS-FGF. Our results indicate that three different EC cell lines, F9, PC-13 and NT2/D1, express transcripts for KS-FGF and that the steady-state levels of these transcripts decrease when EC cells differentiate. Thus, it appears that the KS-FGF gene codes for the FGF-related growth factor produced by EC cells. These findings also suggest a role for KS-FGF during early mammalian development and this possibility is supported by our finding that cultured mouse blastocysts produce growth factors that exhibit the biological properties of an FGF-like growth factor.

INTRODUCTION

During the past decade, EC cells have been used as a model system to identify growth factors that are likely to be produced during early mammalian development and to

---

[1] This work was supported by grants from the National Institutes of Health (HD 21568, HD 19837) and core grants from the National Cancer Institute (CA 36727) and the American Cancer Society (SIG-16).
[2] To whom all correspondence should be sent.

suggest the possible developmental roles of these growth factors (1). The finding that growth of EC cells in serum-free medium is density-dependent (2,3) led to the proposal that EC cells produce several different growth factors (4). Efforts to identify these growth factors led to the finding that EC cells (5) and early mouse embryos (6) produce growth factors that behave as transforming growth factors (TGFs). During the characterization of these growth factors, it was established that EC cells produce a growth factor closely related to platelet-derived growth factor (PDGF) (7,8) and that the production of this growth factor is greatly reduced when EC cells differentiate (7). This factor is immunologically related to PDGF (7) and recent studies suggest that it is likely to be a homodimer of the PDGF A-chain (9,10). Furthermore, it is very likely that this factor is responsible for a significant amount of the TGF activity produced by both EC cells and early mouse embryos (1). Consistent with the latter possibility is the recent finding that early mouse embryos produce transcripts for the PDGF A-chain (11).

EC cells also produce several other growth factors, including: growth factors related to fibroblast growth factor (FGF), insulin-like growth factors, TGF-β, and several less well characterized growth factors (10,12-14). Initial characterization of the FGF-related growth factor established that it is a cationic protein with a molecular weight of approximately 17.5 kDa (12). More recently, we have determined that mouse and human cells produce a heat-labile, heparin-binding growth factor that competes with FGF for binding to membrane receptors (15).

To further our understanding of the FGF-related growth factor produced by EC cells, three questions have been examined. 1) Which form of FGF is produced by EC cells? Currently, five different forms of FGF are known to exist (16-20). 2) How does differentiation of EC cells affect the production of the FGF-related growth factor? 3) Do early mammalian embryos also produce FGF-related growth factors? The work presented in the report argues that EC cells express KS-FGF. This form of FGF, which is also known as KS (21) and hst (18), has been shown to be expressed by a number of different tumor cells (18,22), including human germ-line tumors (23). Our work also argues that differentiation of EC cells reduces the expression of KS-FGF. Lastly, our results with embryos suggest that KS-FGF may be expressed during early development.

METHODS

Cell Culture.

Stock cultures of EC cell lines and SK-HEP-1 cells were maintained as described previously (5,9,15). The mouse and human EC cells were induced to differentiate with retinoic acid using published protocols (3,5,15). NR-6-R cells were maintained and used in the bioassay for FGF as reported previously (24). Cell extracts that were assayed for FGF activity were prepared as described previously (15). Mouse blastocysts from CF-1 females mated with Eppley Swiss males were collected in Standard Egg Culture medium and cultured in a serum-free medium (6). The serum-free medium was a 1:1 mixture of NCTC-109 and enriched Standard Egg Culture medium supplemented with insulin, transferrin, high density lipoprotein and fibronectin. The preparation, storage and use of this medium has been described elsewhere (1).

Northern Blot Analysis.

Northern blot analysis of polyA$^+$ RNA from EC cells and polyA$^+$ RNA from their differentiated cells was performed using standard protocols (9). Northern blot analyses with the cDNA KS-FGF probe were performed by hybridizing the blots at 42°C in 50% formamide and 5X SSPE. These blots were subjected to a final wash in 0.5X SSC at 65°C. The probe for human KS-FGF was a BamHI/EcoRI fragment isolated from the plasmid pG3(B)-SacI (21). Northern blot analyses with an RNA probe for basic FGF (FGFb) were performed by hybridizing the blots at 55°C in 5X SSPE and 40% formamide. These blots were subjected to a final wash in 0.5X SSC at 65°C. The probe for human FGFb was an RNA probe transcribed from the vector pBShbFGF-1 (9).

RESULTS

Differentiation of EC Cells Regulates Production of an FGF-Related Growth Factor.

The levels of the FGF-related growth factor produced by EC cells and by their differentiated cells can be monitored with a bioassay. This assay is performed with a non-transformed indicator cell line, NR-6-R, that only forms soft agar colonies in the presence of FGF or PDGF (24). In this bioassay, PDGF and FGF can be distinguished by their thermal stabilities. PDGF is stable at temperatures above 90°C, whereas FGF is inactivated in less than 10 min at 70°C. The only other growth factor able to influence the soft agar of NR-6-R cells is TGF-β, and TGF-β only potentiates the responses of these cells to FGF (24) and PDGF (Kelly and Rizzino, unpublished data). On its own, TGF-β induces virtually no soft agar growth of NR-6-R cells, whereas 10 pg/ml of FGF induces growth of these cells in soft agar. Thus, NR-6-R serves as a very sensitive indicator cell line for the detection of FGF.

Examination of cell extracts and conditioned media from EC cells and from their differentiated cells indicated that the levels of FGF-related growth factor decrease significantly after EC cells differentiate into parietal endoderm-like cells (15). The largest reductions were observed with mouse F9 EC cells. Within 2 days after treatment with retinoic acid, cell extracts exhibited more than 95% reduction in the levels of the FGF-related growth factor (Table 1). The levels of the FGF-related growth factor also decreased significantly when mouse PC-13 EC cells and human NT2/D1 EC cells differentiated (15).

TABLE 1
EFFECT OF DIFFERENTIATION ON PRODUCTION OF
THE FGF-LIKE GROWTH FACTOR

Cell extract added	Relative FGF activity[a]
F9 EC cells	1.00
Day 2 F9-differentiated cells	0.02
Day 4 F9-differentiated cells	0.05
Day 6 F9-differentiated cells	0.08

[a] The growth-promoting activities of the cell extracts were determined as described previously (24). In each case, the extract from 2×10^6 cells was tested.

Expression of Genes for the FGF Family of Growth Factors.

To identify the gene that codes for the FGF-related growth factor produced by EC cells, we performed northern blot analysis using probes for five different members of the FGF family of growth factors. Although each of the five members of this growth factor family are related to one another at both the amino acid level and the nucleotide level (Table 2), they are sufficiently different from one another to permit identification by northern blot analysis of the gene that codes for the FGF-related growth factor.

TABLE 2
MAXIMAL SIMILARITY OF FGFs AT THE
NUCLEOTIDE[a] AND AMINO ACID[b] LEVELS

	FGFb					Reference
human FGFb	100 (100)	FGFa				(16)
human FGFa	53 (55)	100 (100)	KS-FGF			(17)
human KS-FGF	40 (39)	69 (33)	100 (100)	int-2		(18)
mouse int-2	26 (42)	43 (34)	49 (32)	100 (100)	FGF-5	(19)
human FGF-5	30 (43)	38 (40)	44 (50)	35 (48)	100 (100)	(20)

[a] Percent nucleotide similarity was determined by optimized alignment of coding regions using the NUCALN program (25).
[b] Percent amino acid similarity is shown in brackets as determined by optimized alignment of amino acid sequence. These values were determined using the PROTALN program (25).

Northern blot analysis was performed on polyA$^+$ RNA isolated from three different EC cell lines: mouse F9 EC cells, mouse PC-13 EC cells and human NT2/D1 EC cells. We determined that each of these EC cell lines expresses a 3 kb transcript for KS-FGF (Table 3). In contrast, northern blot analysis of RNA from F9 and PC-13 EC cells did not detect transcripts for FGFb, FGF-5, int-2, or acidic FGF (FGFa). Thus, it appears that KS-FGF is primarily, if not exclusively, responsible for the FGF-related growth factor produced by these EC cells.

Interestingly, the human EC cell line NT2/D1 expresses transcripts that hybridize to a probe for FGFb and the sizes of these transcripts (1.4, 5.1 and 7 kb) are similar to those observed in the human hepatoma cell line SK-HEP-1, which is known to produce FGFb (16). At present, it is unclear why human and mouse EC cells appear to differ in their expression of the FGFb gene, but several possible reasons are considered in the Discussion.

TABLE 3
SIZES OF TRANSCRIPTS THAT HYBRIDIZE WITH PROBES
FOR KS-FGF AND FGFb

Probe	Cells	Transcript sizes (kb)
KS-FGF	NT2/D1	3
	NT2/D1-diff	not detectable
	F9	3
	F9-diff	not detectable
	PC-13	3
	PC-13-diff	not detectable
	SK-HEP-1	not detectable
FGFb	NT2/D1	7.0, 5.1, 1.4
	SK-HEP-1	7.0, 5.1, 1.4
	F9	not detectable

To assess the effect of differentiation on KS-FGF, we examined the expression of this gene by the EC-derived differentiated cells. In the case of mouse EC cells, differentiation was induced by a two-day treatment with 5 μM retinoic acid. This treatment has been shown to induce F9 EC cells to differentiate into cells that exhibit the properties of parietal extraembryonic endoderm. The human

NT2/D1 EC cells were induced to differentiate by a six-day treatment with 1 μM retinoic acid. Our northern blot analysis demonstrated that differentiation dramatically reduces the steady-state levels of KS-FGF transcripts. For each of the differentiated cell populations, we could not detect the presence of the 3 kb KS-FGF transcript. Currently, it is unknown whether this is due to a reduction in transcription and/or a reduction in transcript stability. This question is under investigation.

Conditions for Detecting the Production of Growth Factors by Early Mouse Embryos.

Since at least three EC cell lines produce an FGF-related growth factor, mouse blastocysts were examined for the production of a similar growth factor. To avoid possible contamination with FGF in serum, the blastocysts were cultured in a serum-free medium that contains insulin, transferrin, fibronectin, bovine serum albumin and high density lipoprotein in place of serum. This medium supports attachment of the embryo and outgrowth of the trophoblast, but the cells of the inner cell mass undergo little or no differentiation (1,6). In addition, it is noteworthy that omission of fibronectin prevents attachment of the embryo, whereas omission of high density lipoprotein from this serum-free medium permits attachment, but not outgrowth of the embryos. Examination of attachment and outgrowth in this medium demonstrates that fibronectin is necessary, but not sufficient for outgrowth (1). In the complete serum-free medium, the rate and extent of outgrowth were about 60% of that observed in serum-containing medium (Rizzino and Kelly, manuscript in preparation).

EC cells produce relatively small amounts of the FGF-related growth factor. Therefore, mouse blastocysts were cultured en masse and were bioassayed for the presence of growth factors using the NR-6-R cell bioassay described above. This bioassay was selected for three reasons. It is performed in serum-free medium, it is very sensitive, and, as discussed above, it is specific for FGF and PDGF. (Currently, an equally sensitive, but more specific bioassay for FGF is unavailable.)

Production of Growth Factors by Early Mouse Embryos.

To maximize the detection of growth factors, the embryos were co-cultured with NR-6-R cells after attachment and outgrowth of the embryos had occurred. In these experiments, the embryos were overlaid with a layer of agar, which was overlaid by a second layer of agar containing the NR-6-R cells. Thus, the indicator cells were physically separated from the embryos. If the embryos released FGF- and/or PDGF-related growth factors, they would diffuse through the agar and induce the soft agar growth of the NR-6-R cells. Our results indicate that the embryos released one or more growth factors that induce the soft agar growth of NR-6-R cells (Table 4). Interestingly, addition of TGF-β to the medium potentiated the effect of the embryonic growth factors.

TABLE 4
RELEASE OF A GROWTH FACTOR BY EARLY MOUSE EMBRYOS
THAT INDUCES THE SOFT AGAR GROWTH OF NR-6-R CELLS

Factors added	Number of colonies[a]
None	5
Embryos[b]	90
Embryos + TGF-β (1 ng/ml)	240
TGF-β (1 ng/ml)	15
FGFb (20 pg/ml)	60
FGFb (50 pg/ml)	210
FGFb (150 pg/ml)	545
FGFb (20 pg/ml) + TGF-β (1 ng/ml)	260
FGFb (50 pg/ml) + TGF-β (1 ng/ml)	445

[a] On day 8, the number of NR-6-R cell colonies that formed were determined as described previously (24).
[b] Mouse blastocysts were cultured in serum-free medium. On day 3, the medium was removed from the embryos and a soft agar assay with NR-6-R cells was performed.

These results strongly suggest that early mouse embryos produce FGF- and/or PDGF-related growth factors.

It is possible that a PDGF-related growth factor alone is responsible for the soft agar growth of the NR-6-R indicator cells, since our work and the work of others suggest that early embryonic cells produce a PDGF-related growth factor. Medium conditioned by cultured mouse embryos was also able to stimulate the soft agar growth of NR-6-R cells. Therefore, we attempted to test for the presence of an FGF-related growth factor with an antibody that neutralizes FGFb. (This antibody does not neutralize FGFa, but it may neutralize other forms of FGF, including KS-FGF). We determined that approximately half of the growth factor activity released by the embryos can be neutralized with this antibody, whereas antibodies prepared against acidic FGF do not neutralize the growth factors produced by early embryos (Kelly and Rizzino, unpublished results). Thus, it appears that early embryonic cells produce growth factors belonging to the FGF family of growth factors.

DISCUSSION

The findings described in this report confirm and extend the recent finding that F9 EC cells express the KS-FGF gene (26). In addition to F9 EC cells, we have determined that KS-FGF is expressed in two other EC cell lines. This suggests that most, if not all, EC cell lines express KS-FGF. Our results lead to two other important conclusions. First, the KS-FGF gene is likely to be primarily, if not exclusively, responsible for the FGF-related growth factor produced by EC cells. This conclusion is consistent with our failure to detect transcripts for FGFb, FGFa, FGF-5 or int-2 in RNA from F9 and PC-13 EC cells. Second, the reduction in the production of the FGF-related growth factor that occurs when EC cells differentiate is due to a reduction in the steady-state levels of KS-FGF transcripts.

Although int-2 was not detected on our northern blots, it is reported to be expressed at very low levels by F9 EC cells (less than 1 transcript/cell - 27). Interestingly, the steady-state levels of int-2 dramatically increase when differentiation to parietal endoderm-like cells occurs (27). Although an int-2 protein has not been identified, it may exhibit at least some of the biological properties of FGF. If this is the case, then the differentiation of EC cells could serve as a signal to repress the production of one member of the FGF family (KS-FGF) and induce the expression of another member of this growth factor family. However, if this is the case, the int-2 gene product may only be expressed at very low levels by the differentiated cells observed in our cultures, since relatively little FGF growth factor activity is detected in conditioned medium or cell extracts from EC-derived differentiated cells (15). Alternatively, accurate measurements of the biological activity of the int-2 gene product can not be made with the bioassay used to monitor the activity of the FGF-related growth factor produced by EC cells.

Human EC cells, in apparent contrast to mouse EC cells, not only express the KS-FGF gene, they also express transcripts that hybridize to an FGFb RNA probe. Our failure to detect similar transcripts in F9 EC cells could be due to several different reasons. First, this difference could reflect either a qualitative or a large quantitative difference in the expression of the FGFb gene by human and mouse EC cells. Second, the nucleotide sequence of mouse and human FGFb transcripts may be significantly different, such that our human FGFb probe can not hybridize stably with mouse FGFb transcripts under conditions of moderate to high stringency. Third, the presence of FGFb transcripts could be due to the expression of this gene by differentiated cells that form spontaneously in cultures of NT2/D1 cells. In regard to the last possibility, it is noteworthy that F9 EC cells exhibit very little spontaneous differentiation. At present, we can not eliminate any of these possible explanations, but we favor the last explanation and experiments are in progress to resolve this issue.

The finding that the KS-FGF gene is expressed by EC cells raises the possibility that this gene accounts for some of the FGF activity detected in our embryo cultures. This is very likely given the recent finding that mouse embryos from the eight-cell stage to the blastocyst stage express transcripts for KS-FGF (28). Consequently, we anticipate finding that the KS-FGF gene is expressed by the cells of the inner cell mass, but not by some of its early embryonic derivatives, in particular parietal extraembryonic endoderm.

Currently, the roles of the FGF-related growth factor produced by EC cells and by early embryos are unclear. We have shown previously that differentiation of murine EC cells increases the number of FGF receptors from approximately 2,000 receptors per cell to approximately

17,000 receptors per cell (15). In addition, FGF stimulates the growth of the differentiated cells (15) and, although FGF does not appear to stimulate the growth of murine EC cells (15), it may stimulate the growth of some human EC cells (29). Thus, during early development, it is likely that one or more growth factors belonging to the FGF family regulates growth by paracrine, and possibly autocrine, mechanisms. In addition, since FGF affects the differentiation of many other systems, including mesoderm formation in the frog (30,31), one can anticipate that growth factors belonging to the FGF family are likely to influence both growth and differentiation during early development.

ACKNOWLEDGMENTS

Claudio Basilico is thanked for plasmid pG3(B)-SacI and Heather Rizzino is thanked for editorial assistance.

REFERENCES

1. Rizzino A (1987). Defining the roles of growth factors during early mammalian development. In Bavister BD (ed): "The mammalian preimplantation embryo: Regulation of growth and differentiation in vitro," New York: Plenum Press, p 151.
2. Rizzino A, Sato G (1978). Growth of embryonal carcinoma cells in serum-free medium. Proc Natl Acad Sci USA 75:1844.
3. Rizzino A, Crowley C (1980). Growth and differentiation of embryonal carcinoma cell line F9 in defined media. Proc Natl Acad Sci USA 77:457.
4. Rizzino A, Terranova V, Rohrbach D, Crowley C, Rizzino H (1980). The effects of laminin on the growth and differentiation of embryonal carcinoma cells in defined media. J Supramol Struc 13:243.
5. Rizzino A, Orme L, De Larco JE (1983). Embryonal carcinoma cell growth and differentiation: production of and response to molecules with transforming growth factor activity. Exp Cell Res 143:143.
6. Rizzino A (1985). Early mouse embryos produce and release factors with transforming growth factor activity, In Vitro Cell Dev Biol 21:531.
7. Rizzino A, Bowen-Pope D (1985). Production of PDGF-

like growth factors by embryonal carcinoma cells and binding of PDGF to their endoderm-like differentiated cells. Dev Biol 110:15

8. Gudas LJ, Singh JP, Stiles CD (1983). Secretion of growth regulatory molecules by teratocarcinoma stem cells. In Silver LM, Martin GR, Strickland S (eds): "Teratocarcinoma Stem Cells, Cold Spring Harbor Conferences on Cell Proliferation," Vol. 10. Cold Spring Harbor, NY: Cold Spring Harbor Laboratory, p 229.
9. Tiesman J, Meyer A, Hines RN, Rizzino A (1988). Production of growth factors related to fibroblast growth factor and platelet-derived growth factor by human embryonal carcinoma cells. In Vitro Cell Dev Biol 24:1209.
10. Weima SM, van Rooijen MA, Mummery CL, Feijen A, Kruijer W, de Laat SW, and van Zoelen EJJ (1988). Differentially regulated production of platelet-derived growth factor and of transforming growth factor beta by a human teratocarcinoma cell line. Differentiation 38:203.
11. Rappolee DA, Brenner CA, Schultz R, Mark D, Werb Z (1988). Developmental expression of PDGF, TGF-α, and TGF-β genes in preimplantation mouse embryos. Science 241:1823.
12. Heath JK, Isacke CM (1984). PC13 embryonal carcinoma-derived growth factor. EMBO J 3:2957.
13. Heath JK, Shi W-K (1986). Developmentally regulated expression of insulin-like growth factors by differentiated murine teratocarcinomas and extraembryonic mesoderm. J Embryol Exp Morph 95:193.
14. Jakobovits A, Banda MJ, Martin GR (1985). Embryonal carcinoma-derived growth factors: Specific growth-promoting and differentiation-inhibiting activities. In Feramisco J, Ozanne B, Stiles C (eds): "Cancer Cells: Growth Factors and Transformation," Vol. 3. Cold Spring Harbor, NY: Cold Spring Harbor Laboratory, p 393.
15. Rizzino A, Kuszynski C, Ruff E, Tiesman J (1988). Production and utilization of growth factors related to fibroblast growth factor by embryonal carcinoma cells and their differentiated cells. Dev Biol 129:61.
16. Abraham JA, Whang JL, Tumolo A, Mergia A, Friedman J, Gospodarowicz D, Fiddes JC (1986). Human basic fibroblast growth factor: Nucleotide sequence and genomic organization. EMBO J 5:2523.

17. Jaye M, Howk R, Burgess W, Ricca GA, Chiu M-I, Ravera MW, O'Brien SJ, Modi WS, Maciag T, Drohan WN (1986). Human endothelial cell growth factor: Cloning, nucleotide sequence, and chromosomal localization. Science 233:541.
18. Taira M, Yoshida T, Miyagawa K, Sakamoto H, Terada M, Sugimura T (1987). cDNA sequence of human transforming gene hst and identification of the coding sequence required for transforming activity. Proc Natl Acad Sci USA 84:2980.
19. Moore R, Casey G, Brookes S, Dixon M, Peters G, Dickson C (1986). Sequence, topography and protein coding potential of mouse int-2: A putative oncogene activated by mouse mammary tumour virus. EMBO J 5:919.
20. Zhan X, Bates B, Hu X, Goldfarb M (1988). The human FGF-5 oncogene encodes a novel protein related to fibroblast growth factors. Mol Cell Biol 8:3487.
21. Delli Bovi P, Curatola AM, Kern FG, Greco A, Ittman M, Basilico C (1987). An oncogene isolated by transfection of Kaposi's sarcoma DNA encodes a growth factor that is a member of the FGF family. Cell 50:729.
22. Tsutsumi M, Sakamoto H, Yoshida T, Kakizoe T, Koiso K, Sugimura T, Terada M (1988). Coamplification of the hst-I and the int-2 Genes in Human Cancers. Jpn J Cancer Res (Gann) 79:428.
23. Yoshida T, Tsutsumi M, Sakamoto H, Miyagawa K, Teshima S, Sugimura T, Terada M (1988). Expression of the HST1 Oncogene in Human Germ Cell Tumors. Biochem Biophys Res Commun 155:1324.
24. Rizzino A, Ruff E (1986). Fibroblast growth factor induces the soft agar growth of two non-transformed cell lines. In Vitro Cell Dev Biol 22:749.
25. Wilbur WJ, Lipman DJ (1983). Rapid similarity searches of nucleic acid and protein databanks. Proc Natl Acad Sci USA 80:726.
26. Yoshida T, Muramatsu H, Muramatus T, Sakamoto H, Katoh O, Sugimura T, Terada M (1988). Differential Expression of Two Homologus and Clustered Oncogenes, Hst1 and Int-2, During Differentiation of F9 Cells. Biochem Biophys Res Commun 157:618.
27. Jakobovits A, Shackleford GM, Varmus HE, Martin GR (1986). Two proto-oncogenes implicated in mammary carcinogenesis, int-1 and int-2, are independently regulated during mouse development. Proc Natl Acad Sci USA 83:7806.

28. Rappolee DA, Schultz GA, Pedersen RA, Werb Z (1988). Expression of Genes for Growth Factors and Growth Factor Receptors in Preimplantation Mouse Embryos. J Cell Biol 107:234a.
29. Graham CF. personal communication.
30. Slack JMW, Darlington BG, Heath JK, Godsave SF (1987). Mesoderm induction in early Xenopus embryos by heparin-binding growth factors. Nature 326:197.
31. Kimmelman D, Kirshner M (1987). Synergistic induction of mesoderm by FGF and TGF-β and the identification of an mRNA coding for FGF in the early Xenopus embryo. Cell 51:869.

THE ENERGY METABOLISM OF THE PREIMPLANTATION EMBRYO

Henry J. Leese[1]

Department of Biology, University of York,
Heslington, York YO1 5DD, U.K.

ABSTRACT The results of non-invasive studies on single mouse embryos and a review of related literature are used to derive a model of preimplantation embryo metabolism. It is proposed that prior to compaction, mouse embryos adopt an aerobic type of metabolism, oxidising pyruvate, lactate, amino acids or possibly fatty acids. Excessive utilisation of glucose during this period is potentially deleterious. Around the time of compaction, the embryos switch to a metabolism based on glucose, which is used as a source of energy, to provide precursors for biosynthetic reactions and to prime the embryos for the anoxic environment they will encounter at implantation.

ENERGY METABOLISM OF THE PREIMPLANTATION EMBRYO

Most of the experimental results to be presented were obtained using a non-invasive assay technique devised to measure nutrient uptake by single preimplantation mouse embryos (1).

This method was based on the ultramicrofluorometric technique devised by Mroz & Lechene (2) to measure picomole quantities of urea. In this technique, urea assays were carried out in siliconized capillary tubes and the reaction product quantified using a fluorescence microscope. The

---

[1]The author's studies reported in this paper were largely supported by the U.K. Science and Engineering Research Council.

reagents and samples were delivered using micropipettes in the picoliter and nanoliter range made on a microforge (2).

Leese et al (3) modified the method by carrying out reactions in nanoliter-sized droplets on siliconized microscope slides under a layer of mineral oil, rather than in capillary tubes. The use of microdroplets made for easier handling of samples and reagents and enabled extra reagent(s) to be added to the droplets if necessary. Working with microdrops, it was necessary to increase the concentration of the coupling enzyme(s) used in the reaction assay in order to overcome the denaturation of enzymes at the interface between the microdrops and the mineral oil (3).

Leese et al (3) further modified the method to permit measurement of the whole range of metabolites conventionally assayed in coupled enzymatic reactions involving the generation or consumption of the pyridine nucleotides NADH and NADPH (4). The specific compounds measured to begin with were ATP, ADP and AMP.

It was found that single mouse oocytes had a high ATP content relative to ADP and a high energy charge (0.83). As development progressed, the ADP content fell, that of ADP rose and AMP remained unchanged so that the energy charge dropped at the 8-cell and morula stages to 0.62 and 0.58 respectively. These changes were consistent with the relatively low respiratory rates observed at the early stages of development and a stimulation in oxygen uptake at the onset of compaction (5).

Leese & Barton (1) modified this technique to measure the uptake of glucose and pyruvate by single or small numbers of LACA-strain mouse oocytes, zygotes, 2, 4 and 8-16-cell embryos, morulae and blastocysts. The embryos were incubated for up to 3 hours in 20 nl medium M2 containing 0.33 mM pyruvate, 1 mM glucose and 23.3 mM lactate. The pyruvate and glucose concentrations are close to those found in rabbit oviduct fluid (6). Serial samples in the range 400 pl - 4 nl were removed at 30 min intervals and analysed for their nutrient content. The data (Figure 1) indicated that unfertilized and fertilized oocytes and embryos up to the 8-16 cell stage have a preference for pyruvate before switching to glucose in the blastocyst. The embryos in this study were discarded after their metabolism had been assessed.

Gardner & Leese (7) extended this work by taking single preimplantation embryos at the 2-cell stage, measuring their nutrition and then culturing them overnight in Medium M16. Metabolism was re-assessed the following day and so on until the blastocyst stage was reached. These embryos were the

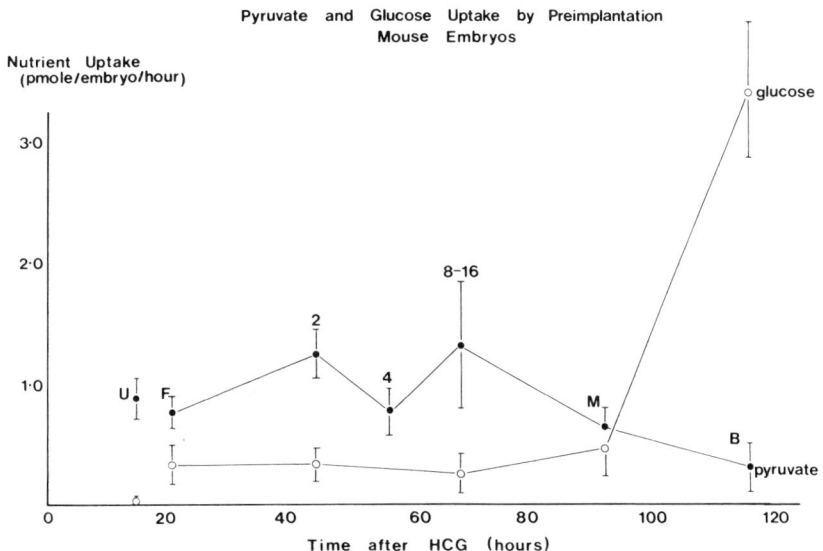

FIGURE 1. The uptake of pyruvate and glucose by unfertilized (U), and fertilized (F) mouse ova, and the 2-cell (2) and 4-cell (4), 8-16-cell (8-16), morula (M) and blastocyst (B) stages of development. Values are mean ± s.e.m. "Journal Of Reproduction and Fertility 72, 9 13 (1984) Reproduced with permission."

progeny of F1 females (CBA/Ca * C57BL/6) mated with F1 males. They were incubated in 25 nl M2 containing 0.33 mM pyruvate and 1 mM glucose and 23.3 mM lactate.

Although the pattern of the results was similar to that obtained with embryos removed at discrete stages, they differed in that the transition from pyruvate to glucose at the late morula stage was more striking. It was also notable that embryos cultured from the 1-cell stage were retarded and failed to exhibit the sharp transition from pyruvate to glucose. The likely explanation for this is that the HEPES buffered M2 was having a deleterious effect on the metabolism of the 1-cell embryos. A similar phenomenon has recently been reported by Butler et al (8) and earlier, by Farrell & Bavister (9). The 1-cell mouse embryo seems to be particularly sensitive to external factors in vitro.

Qualitatively similar results on the pattern of energy substrate uptake have been obtained for the mouse and rabbit by Brinster (10) using isotopically-labelled nutrients, by Wales (11), also for the mouse, and by Flood & Wiebold (12) for glucose uptake by pig preimplantation embryos. Rieger & Guay (13) have shown that cattle embryos exhibit a somewhat different pattern.

The Nature of the Block to Glucose Utilisation in Early Mouse Embryos

Using an enzymatic cycling technique on single mouse embryos, Barbehenn et al (14, 15) obtained evidence, summmarised by Biggers et al (16) which suggested that glucose utilisation during the early preimplantation stages was blocked at the level of the glycolytic enzyme phosphofructokinase.

The extent to which changes in the transport of pyruvate and glucose across the embryo plasma membrane could play a role in regulating nutrient uptake has recently been examined by Gardner & Leese (17) using the non-invasive approach outlined above. A glucose facilitated diffusion system was identified in single mouse blastocysts. The system exhibited substrate saturation kinetics with an apparent Kt for the carrier-mediated component of 0.14 mM. Using phloretin, a specific inhibitor of glucose facilitated diffusion, it was demonstrated that the carrier-mediated component was present from the 2-cell stage onwards (i.e. when glucose uptake is first detectable). Although this would appear to rule out a role for glucose transport in limiting glucose uptake during the early developmental stages, such a conclusion has to be tentative since the intracellular glucose concentration, which would need to be known to establish this unequivocally, has not been determined.

Hooper & Leese (18) determined the activity of hexokinase in mouse oocytes and preimplantation embryos (Figure 2). At each stage of development, the activity was about 4-fold greater than the rate at which glucose is consumed in culture suggesting that hexokinase alone is not responsible for regulating glucose uptake. However, the enzyme activity increased dramatically at the morula stage, coincident with the increase in glucose uptake, and the likelihood that the enzyme is inhibited to some extent in the cell leaves open the possibility of a regulatory role.

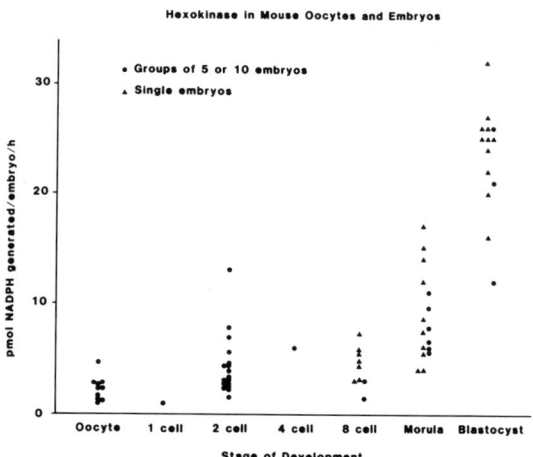

FIGURE 2. The activity of hexokinase in mouse oocytes and preimplantation embryos. "Biochemical Society Transactions. In Press. Copyright (C) 1989. The Biochemical Society, London.

A carrier-mediated pyruvate entry system was demonstrated in unfertilized mouse oocytes by Leese & Barton (1) on the basis of the inhibition of pyruvate uptake by the inhibitor α-cyano-4-hydroxycinnamate. Gardner & Leese (17) showed that this component was present throughout preimplantation development indicating that the switch from a pyruvate to a glucose-based metabolism at the late morula stage is not due to the disappearance of a carrier-mediated pyruvate entry system. Furthermore, with pyruvate as the sole energy substrate in the medium used to assess metabolism, the uptake values were almost 2-fold greater than in the presence of glucose or lactate. These data, which illustrate the plasticity in nutrient preference by early mouse embryos, may be summarised as follows:

1) Preimplantation mouse embryos presented with a physiological mixture of nutrients normally switch from pyruvate to glucose as the preferred substrate around the late morula stage.

2) In the absence of glucose, mouse embryos can continue to consume pyruvate at high levels.

3) A high lactate concentration "damps down" pyruvate uptake.

The Formation of Lactate

Gardner & Leese (17) reported that 25% of the glucose consumed by single 2-cell embryos could be accounted for by lactate formation; the figure for the blastocyst stage was 44%. It should be noted that these values represent an "aerobic glycolysis" since the embryos were incubated in an atmosphere containing 20% oxygen. The question therefore arises as to why the mouse embryo should metabolise a carbon source in this manner, when the complete oxidation of glucose to CO_2 and H_2O would provide approximately 18 times more ATP/mole.

Clues to the reason for this phenomenon come from three sources: (i) an examination of the glucose metabolism of rodent embryos during the peri-implantation period; (ii) a consideration of the roles of glucose other than as an energy source and (iii) the behaviour of other mammalian cell types in culture.

The Glucose Metabolism of Peri-implantation Rodent Embryos

The energy metabolism of the rodent embryo at implantation has recently been reviewed by Leese (19) and Gardner (20). Briefly, Clough & Whittingham (21) showed that mouse embryos removed from the uterus on days 6-9 post-coitum convert more than 90% of the glucose they consume to lactate even under aerobic conditions. Ellington (22) has recently shown that the energy metabolism of the rat embryo between 9.5 and 10.5 days is also characterized by high rates of glycolysis with 100% of the glucose taken up accounted for by lactate formation. Leese (19) proposed an explanation for this high quantitative conversion of glucose to lactate in terms of the microvasculature of the rodent uterus at implantation, where Rogers and co-workers (23,24), using the technique of microvascular casting were able to produce a 3-dimensional replica of the vasculature immediately surrounding the implanting rat blastocyst on day 6 of pregnancy. It was found that the centre of the implantation sites were characterised by the absence of patent capillaries in an area approximately 420 μ long by 210 μ in diameter. The size and location of this space corresponded to the primary decidual zone seen in light microscope sections taken on day 6 of pregnancy. In other words, the implanting rodent blastocyst is surrounded by a region largely devoid of capillaries, which would explain its dependence on anaerobic glycolysis to obtain its supply of ATP.

The increasing formation of lactate by the preimplantation mouse embryo may be seen as a preparation for the anoxic conditions to be encountered in the uterus at implantation.

The Requirement of Preimplantation Embryos for Glucose

Glucose appears to be required in increasing quantities by the mouse embryo from the late morula stage onwards. This is a time of great metabolic activity as the embryonic genome becomes fully activated and the process of compaction and the need to create the blastocoel cavity impose a heavy demand for ATP. In addition, glucose is a precursor of many, particularly macromolecular cell constitutents; ribose moieties for nucleic acid biosynthesis; glycerol phosphate for the formation of phospholipids; complex sugars in mucoproteins and mucopolysaccharides (19,20). Glucose may also provide the carbon moieties for non-essential amino acids. Prior to the time of compaction, glucose appears to be less essential to the early embryo; indeed as is extensively documented elsewhere in this volume, it is at best "tolerated" and at worst, "inhibitory" (Bavister; personal communication and (25), (26)).

The Glucose Metabolism of Mammalian Cells in Culture

Otto Warburg (27) first showed that the growth of certain cells in culture is characterised by a decline in oxidative metabolism and an increase in aerobic glycolysis. For example, hetpatocytes when first placed in culture will consume lactate but within 30 minutes begin to produce lactate in considerable quantities (28). This phenomenon, which is also characteristic of cultured tumor cells, has been admirably summarised by Mandel (29).

These considerations and the data presented earlier, may be brought together in a model which accounts for preimplantation energy metabolism (Figure 3).

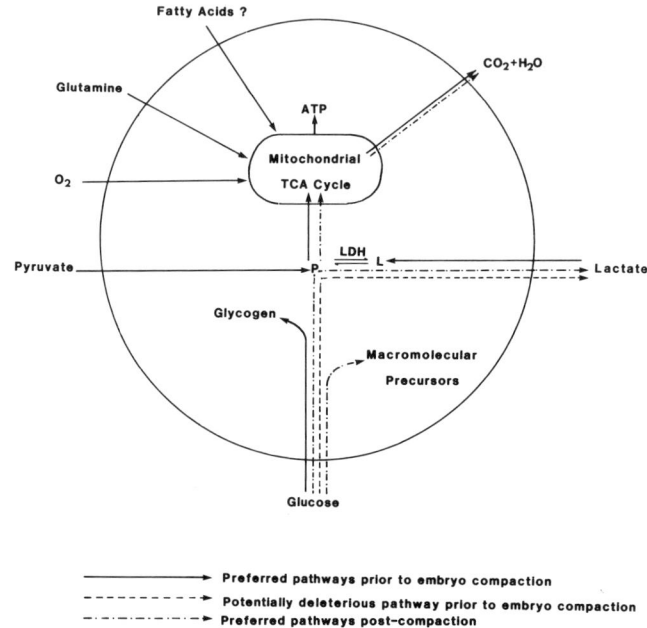

FIGURE 3. Model of preimplantation embryo metabolism.

A basic premise of the model is that during the developmental stages prior to compaction, the embryo prefers to adopt a highly aerobic type of metabolism in order to obtain its supply of ATP. It respires pyruvate, lactate, or if they are available, amino acids such as glutamine. Such a pattern may have evolved from a metabolism based on fatty acids derived from yolk; it is notable that lower vertebrates such as amphibians, reptiles and birds have yolky eggs (30). Furthermore, Kane (31) has shown in the rabbit, that certain fatty acids can support the growth of fertilized 1-cell eggs to morulae. It may also be of significance that one such fatty acid, propionate, may be converted by the rabbit oviduct into pyruvate (32).

During the early preimplantation stages, it would appear that the embryo does not wish to compromise this highly aerobic metabolism by consuming glucose in any quantity, except, perhaps, as a precursor of glycogen (33,34).

The need to consume glucose arises at compaction and blastocoel cavity formation where, in addition to acting as an oxidisable enrgy source, it performs the functions of providing the carbon moieties for macromolecule and non-essential amino acid biosynthesis, and of priming the embryo's glycolytic capacity in anticipation of the anoxic environment it will encounter at implantation.

When early embryos and other types of mammalian cells are placed in culture, they seem, for reasons that are at present unclear, to assume a glycolytic type of metabolism which appears to interfere with their capacity to develop optimally and may lead them to become arrested. Two strategies which can overcome these developmentally arrested states are a very high ratio of lactate to pyruvate and/or the presence of glutamine in the culture medium (26,35). Glutamine is well known as an energy source for cultured mammalian cells (36) and is consumed by rabbit oocytes (37) and early mouse embryos (38). It is readily transported and metabolised and may be acting to counteract the tendency of early embryos in culture to consume glucose.

A high lactate:pyruvate ratio, which is found in many culture media, may perform the same function by forcing the equilibrium reaction catalysed by lactate dehydrogenase over to the pyruvate side and inhibiting the uptake and conversion of glucose to lactate by mass action. The kinetic properties of lactate dehydrogenase are consistent with this proposition; it is present at very high concentrations in early mouse embryos (39) and in all probability brings about an equilibration between pyruvate and lactate within the embryo.

ACKNOWLEDGEMENTS

I thank Susan Heyner and Lynn Wiley for providing an atmosphere at the UCLA Symposium so conducive to the development of ideas, and my colleague David Gardner for skilfully producing much of the data reported in this paper.

REFERENCES

1. Leese HJ, Barton AM (1984). Pyruvate and glucose uptake by mouse ova and preimplantation embryos. J Reprod Fert 72:9.
2. Mroz EA, Lechene C (1980). Fluorescence analysis of picoliter samples. Analyt Biochem 102:90.
3. Leese HJ, Biggers JD, Mroz EA, Lechene C (1984). Nucleotides in a single mammalian ovum or preimplantation embryo. Analyt Biochem 140:443.
4. Bergmeyer H-U, Gawehn K (1974). "Methods of Enzymatic Analysis". 2nd English edn New York: Academic Press.
5. Mills RM, Brinster RL (1967). Oxygen consumption of preimplantation mouse embryos. Exp Cell Res 47:337.
6. Leese HJ, Barton AM (1985). Production of pyruvate by isolated mouse cumulus cells. J. exp Zool 234:241.
7. Gardner DK, Leese HJ (1986). Non-invasive measurement of nutrient uptake by single cultured pre-implantation mouse embryos. Hum Reprod 1:25.
8. Butler JE, Lechene C, Biggers JD (1988). Noninvasive measurement of glucose uptake by two populations of murine embryos. Biol Reprod 39:779.
9. Farrell PS, Bavister BD (1984). Short-term exposure of two-cell hamster embryos to collection media is detrimental to viability. Biol Reprod 31:109.
10. Brinster RL, (1973). Nutrition and metabolism of the ovum, zygote and blastocyst. In Greep RO, Astwood EB (eds) "Handbook of Physiology" Section 7, Vol II part 2 p 165 Washington DC: Am Physiol Soc.
11. Wales RG, (1986). Measurement of metabolic turnover in single mouse embryos. J Reprod Fert 76:717.
12. Flood MR, Wiebold JL (1988). Glucose metabolism by preimplantation pig embryos. J Reprod Fert 84:7.
13. Rieger D, Guay P (1988). Measurement of the metabolism of energy substrates in individual bovine blastocysts. J Reprod Fert 83:585.
14. Barbehenn EK, Wales RG, Lowry OH (1974). The explanation for the blockade of glycolysis in early mouse embryos. Proc natn Acad Sci USA 71:1056.
15. Barbehenn EK, Wales RG, Lowry OH (1978). Measurement of metabolism in single preimplantation embryos; a new means to study metabolic control in early embryos. J Embryol exp Morph 43:29.
16. Biggers JD, Gardner DK, Leese HJ (1989). Control of Carbohydrate Metabolism in Preimplantation Mammalian Embryos. In Rosenblum IY, Heyner S (eds) "Regulation of Growth in Development", CRC Press.

17. Gardner DK, Leese HJ (1988). The role of glucose and pyruvate transport in regulating nutrient utilisation by preimplantation mouse embryos. Development 104:423.
18. Hooper MAK, Leese HJ (1989). The activity of hexokinase in mouse oocytes and preimplantation embryos. Biochem Soc Trans (In Press).
19. Leese HJ (1989). Energy metabolism of the blastocyst and uterus at implantation. In K. Yoshinaga (ed) Monograph on Blastocyst Implantation. Boston, Mass, Adams Publishing.
20. Gardner DK (1987). D. Phil Thesis, University of York, U.K.
21. Clough JR, Whittingham DG (1983). Metabolism of ^{14}C glucose by preimplantation mouse embryos in vitro. J Embryol exp Morph 74:133.
22. Ellington SKL (1987). In vitro analysis of glucose metabolism and embryonic growth in post-implantation rat embryos. Development 100:431.
23. Roger PAW, Murphy CR, Gannon BJ (1982). Changes in the spatial organisation of the uterine vasculature during implantation in the rat. J Reprod Fert 65:211.
24. Roger PAW, Murphy CR, Rogers AW, Gannon BJ (1983). Capillary patency and permeability in the endometrium surrounding the implanting rat blastocyst. Int J Microcirc: Clin Exp 2:241.
25. Bavister BD (1989). Energy substrate regulation of pre-implantation development in vitro. Journal of Cellular Biochemistry Suppl 13B:188.
26. Chatot CL, Ziomet CA (1989). An improved culture medium promotes development of 1-cell mouse embryos in vitro. Journal of Cellular Biochemistry Suppl 13B:195.
27. Warburg O (1926). "Uber den stoffwechsel der tumoren" Berlin Springer Verlag.
28. Gerhardt R, Bellemann P, Mecke D (1978). Metabolic and enzymatic characteristics of adult rat liver parenchymal cells in non-proliferating primary monolayer cultures. Exp Cell Res 112:431.
29. Mandel LJ (1986). Energy metabolism of cellular activation, growth and transformation. Current topics in membranes and transport 27:261.
30. Pedersen RA (1989). In Heyner S, Wiley L (eds) "Early embryo development and paracrine relationships". UCLA Symposium on Molecular and Cell Biology, New Series vol 117. New York Alan R. Liss.
31. Kane MT (1979). Fatty acids as energy sources for culture of one-cell rabbit ova to viable morulae. Biol Reprod 20:323.

32. Leese HJ (1980). The stimulation of pyruvate appearance in the rabbit oviduct lumen by sodium propionate. J Reprod Fert 59:421.
33. Stern S, Biggers JD (1968). Enzymatic estimation of glycogen in the cleaving mouse embryo. J exp Zool 168:61.
34. Pike IL, Wales RG (1982). Uptake and incorporation of glucose especially into the glycogen pools of pre-implantation mouse embryos during culture in vitro. Aust J. Biol Sci 25:195.
35. Carney EW, Bavister BD (1987). Stimulatory and inhibitory effects of amino acids on the development of hamster eight-cell embryos in vitro. Journal of In Vitro Fertilization and Embryo Transfer 4: 162.
36. Zielke HR, Zielke CL, Ozand PT (1984). Glutamine: a major energy source for cultured mammalian cells. Fed Proc 43:121.
37. Bae I-H, Foote RH (1975). Carbohydrate and amino acid requirement and ammonia production of rabbit follicular oocytes matured in vitro. Exp Cell Res 91:113.
38. Gardner DK, Clarke RC, Lechene C, Biggers JD (Unpublished observations).
39. Brinster RL (1965). Lactate dehydrogenase activity in the preimplantation mouse embryo. Biochim biophys Acta 110:439.

REGULATION OF HAMSTER PREIMPLANTATION EMBRYO DEVELOPMENT IN VITRO BY GLUCOSE AND PHOSPHATE[1]

Barry D. Bavister

Department of Veterinary Science, University of Wisconsin, Madison, Wisconsin 53706

ABSTRACT Preimplantation hamster embryos show complete blocks to development at the 2-cell and 4-cell stages. These blocks preclude studies on the viability of in vitro fertilized hamster eggs and impede use of hamster embryos for obtaining comparative data on early developmental regulation. The primary cause of these blocks has now been identified as inorganic phosphate, with secondary inhibition due to glucose. Deletion of these components from the culture medium allows some hamster 2-cell embryos to develop into blastocysts in vitro.

INTRODUCTION

In striking contrast to the very large body of knowledge about the regulation of preimplantation embryo development that has been derived from studies with mouse embryos, comparative information from most other species is sparse. A major reason for this lack of knowledge is the existence of blocks to development shown by embryos of many

---

[1]This work was done as part of the National Cooperative Program on Non-Human In Vitro Fertilization and Preimplantation Development and was funded by the National Institute of Child Health and Human Development, NIH, through cooperative agreement HD-22023.

species under culture conditions. For example, hamster and rat embryos block in vitro between the 2-cell and the 8-cell stages, embryos from many strains of mice undergo only one cleavage division in vitro when cultured from the 1-cell stage, and embryos of domestic species show blocks at the 4-cell or 8- to 16-cell stages, depending on the species (1). Even primate embryos, which do not show overt blocks to development in vitro, nevertheless fail to show normal development in culture, and are mostly unable to establish pregnancies after transfer. Substantial differences must exist between culture conditions that are used for embryos and the natural environment (1). Although the blocks to development in vitro present a major obstacle to progress, they also represent a unique opportunity to gain insights into the requirements of cultured embryos in a wide variety of species. My philosophy is that, once we learn how to overcome the blocks to development in a particular species, the resulting information should give us important clues about the control of embryonic development. In addition, perhaps information derived from one species might have some general relevance that will help to overcome the blocks to development in embryos of other animals.

For the past six years, my laboratory has been studying the ability of hamster embryos to develop in culture, in part because we have used this species for many years for studies on in vitro fertilization, and the existence of the blocks precludes analysis of the developmental capability of in vitro fertilized embryos. Hamster embryos usually show blocks at several stages during in vitro culture. Hamster eggs can readily be fertilized in vitro and many of the zygotes can develop as far as the 2-cell stage, as first shown by Yanagimachi and Chang about 25 years ago (2,3). However, 2-cell hamster embryos derived from eggs fertilized in vitro or in vivo always fail to undergo any further cleavage. Similarly, 4-cell embryos recovered from mated females completely fail to cleave to the 8-cell stage in vitro. Not surprisingly, the golden hamster has not been a popular species for research into preimplantation embryo development!

EARLY STUDIES ON BLOCKS TO DEVELOPMENT IN CULTURED HAMSTER EMBRYOS

In vitro fertilized hamster eggs are not viable, as shown by embryo transfer experiments (4). In vivo fertilized embryos collected at the 1- or 2-cell stages and immediately transferred to recipients also fail to continue development, whereas (non-cultured) 4- and 8-cell embryos are viable following transfer (5, 6). Thus, early cleavage stage hamster embryos are unusually susceptible to exposure to culture media. In 1983, we found that 8-cell hamster embryos would grow to the blastocyst stage in vitro, using a culture medium developed for in vitro fertilization (TALP), provided that certain amino acids were included (7). This requirement showed that hamster embryos differ from the standard model, i.e., mouse embryos, which do not need any exogenous amino acids for development up to the blastocyst stage in vitro (8). The amino acids used to support 8-cell hamster embryo development (7) were those required by hamster oocytes for maturation in vitro: glutamine, isoleucine, methionine and phenylalanine (9). When these amino acids were included in the culture medium, 36% of early (54 hours post egg activation) 8-cell embryos cleaved to the blastocyst stage vs. only 2% in the absence of the amino acids (7). Curiously, when 8-cell hamster embryos were recovered only a few hours later (61 hours post-activation), many more (22%) of these embryos were able to develop into blastocysts without exogenous amino acids, indicating that a substantial shift in nutrient requirements and/or in metabolic capabilities occurs in the hamster embryo at the 8-cell stage.

Using the same culture solution (TALP) supplemented with four amino acids that supported development of 8-cell embryos to the blastocyst stage (7), we were unable to grow either 2-cell or 4-cell hamster embryos in vitro. Since 1983, a variety of approaches has been tried in our laboratory to support development of these early cleavage stages in vitro, including use of complex culture media, sometimes supplemented with serum, but these changes either did not help, or only seemed to make the situation worse by causing cytoplasmic deterioration.

In 1984, we discovered that hamster 2-cell embryos lost their viability within a few minutes of exposure to air-buffered simple balanced salt solutions (10). This result indicated that the medium itself was detrimental to the embryos, instead of the more obvious conclusion that the medium lacked some essential growth factor normally present in vivo. With this knowledge, we constructed two hypotheses: either some component present in the culture medium was detrimental to hamster 2-cell embryos, or some essential factors normally present within the embryos were leaking out into the external medium, preventing further development. A consistent sign of damage in cultured hamster 2-cell embryos was swelling of the blastomeres, which sometimes obliterated the perivitelline space (11). This effect became worse as the concentration of bovine serum albumin (BSA) used in the TALP medium was increased. We replaced the BSA with polyvinylalcohol (PVA), which is widely used in my laboratory for gamete and embryo manipulations (12). The hamster 2-cell embryos appeared more normal when incubated in this chemically-defined medium (TLP-PVA): the blastomere swelling effect was less severe, but still the embryos did not develop.

RECENT DISCOVERIES ABOUT THE CAUSES OF THE EARLY BLOCKS TO DEVELOPMENT

Results Using Microdrop Cultures

Since embryos developing within the oviduct are maintained in very small volumes of fluid, we placed hamster 2-cell embryos into an exceedingly small drop of culture medium (forty embryos in less than 1 μl) so that the embryo: medium volume ratio was greatly increased (13). In addition to the usual four amino acids (7), glycine and/or taurine (1mM each) were also included. Under these conditions, the majority of hamster 2-cell embryos were able to develop to the 3- or 4-cell stage, which we had never observed before under in vitro conditions. Although this microdrop culture system was impractical for routine use, it showed that it was possible to support development of hamster embryos through the 2-cell block. All of our subsequent experiments were done with standard volume (100µl) culture drops under oil incubated at 37°C in 5% to 10% CO_2 in air.

Inhibition of Embryo Development by Glucose

Since the results with 1µl microdrops (13) did not indicate which of our two hypotheses was correct, we assumed that both were partly valid. To compensate for possible leakage of amino acids from the embryos, we increased the number of amino acids in the culture medium to 20 (14). Four of the amino acids (glutamic acid, aspartic acid, taurine and glycine) were each present at 5 to 7 mM, comparable to their concentrations in mouse and rabbit reproductive tract fluids (15).

The first insight into the causes of the blocks to development was obtained when we decided to delete one of the existing components of the culture medium. The rationale for making this change was as follows. Since sperm undergo their final preparations for fertilization within the oviduct, where the first few cleavage divisions of embryos also take place, it seemed possible that sperm, eggs and embryos might have some similarities in their metabolic activities and nutrient requirements. Glucose is usually found in culture media, but it can inhibit fertilizing ability (acrosome reactions) of spermatozoa (16, 17). It seemed possible from the foregoing argument that glucose might also inhibit early embryo development. This prediction proved to be correct: when glucose was deleted from the culture medium, about 23% of 2-cell hamster embryos cleaved to the 4-cell stage after 24 hours of culture (Fig. 1). Although this proportion did not increase significantly after culture for one more day, it was nevertheless a vast improvement on the 0% development of cultured hamster 2-cell embryos that we and others had always previously observed (other than in the 1µl microdrops: 13).

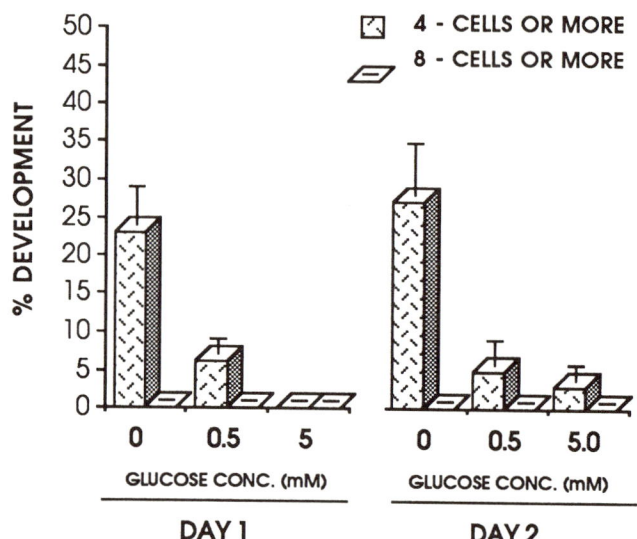

Figure 1. Effect of glucose concentration on development of 2-cell hamster embryos in vitro. Embryos were cultured in chemically-defined medium TLP-PVA with 0.35 mM Pi and 20 amino acids for 21-24 hours (day 1) or 44-46 hours (day 2). Values are mean % ± S.E.M. of 15 trials. Data from (14).

Inhibition of Embryo Development by Phosphate

What could be the explanation for the inhibitory effect of glucose on hamster embryo development? One possibility seemed to be the "Crabtree" effect, which was originally described in 1929 as glucose-mediated inhibition of respiration in mouse ascites tumor cells (18). In this "reverse Pasteur effect," the presence of glucose stimulates glycolysis at the expense of oxidative phosphorylation. As shown much later, the "Crabtree" effect is exacerbated by low concentrations of inorganic phosphate (Pi) in the culture medium: when Pi concentration was raised, the inhibitory effect of glucose was reduced or eliminated (19). The explanation proposed is that Pi stimulates glycogenolysis (via the enzyme phosphorylase) and glycolysis (via hexokinase, phospho-fructokinase and glyceraldehyde-3-phosphate dehydrogenase), and is "consumed" by the phosphorylation reactions (19, 20). The net result would be depletion of intracellular Pi with consequent restriction of the embryo's ability to produce energy by the more efficient mitochondrial pathways (TCA cycle and oxidative phosphorylation).

Now, our culture medium for hamster embryos is based on Tyrode's solution, which has one of the lowest Pi concentrations of any culture medium: only 0.35 mM, compared to greater than 1 mM for most standard culture media (8). It seemed obvious that if we raised the Pi concentration in our culture medium, then we could eliminate the inhibitory effect of glucose. This prediction was totally wrong! When four concentrations of Pi (0, 0.1, 0.35 and 1.05 mM) were tested, only the "low control" medium in which Pi was completely absent supported substantial development of hamster 2-cell embryos. After culture for 2 days in the Pi-free medium, 97% of embryos were at the 4-cell stage or further, and 26% had reached the 8-cell stage. This was the first time in our experience that a significant proportion of 2-cell hamster embryos had completed two cleavage divisions in culture.

The relative roles of glucose and Pi were then tested in a factorial experiment (Fig. 2). The high concentration of glucose (5mM) was selected as representative of the concentration in most culture media and in blood of most mammals; 0.35mM Pi is approximately physiological and is the level in TLP-PVA. In the absence of Pi, regardless of the concentration of glucose, between 55% and 65% of hamster 2-cell embryos developed to the 4-cell stage within the first day of culture, whereas in the presence of Pi even at only 0.1 mM, there was virtually no development.

These experiments showed that glucose was only inhibitory in the presence of Pi, and removing glucose alone from the medium allowed a small but significant amount of development of 2-cell hamster embryos. In contrast, removal of Pi allowed a substantial proportion of hamster embryos to develop to 4- and 8-cells in vitro, but glucose was no longer inhibitory in the absence of Pi.

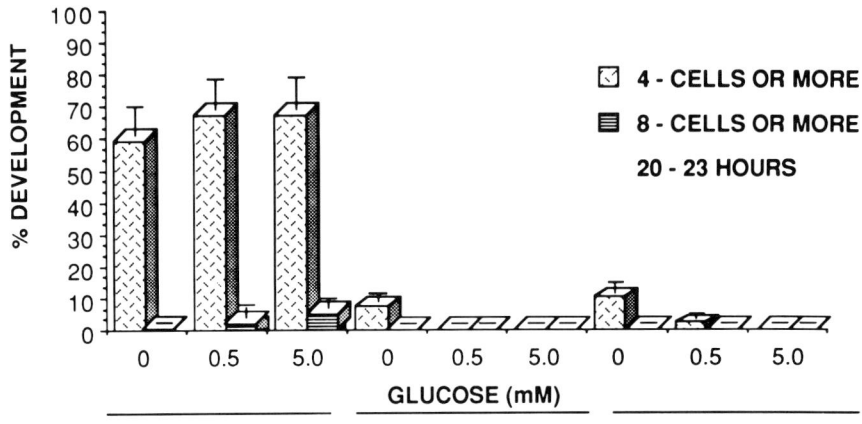
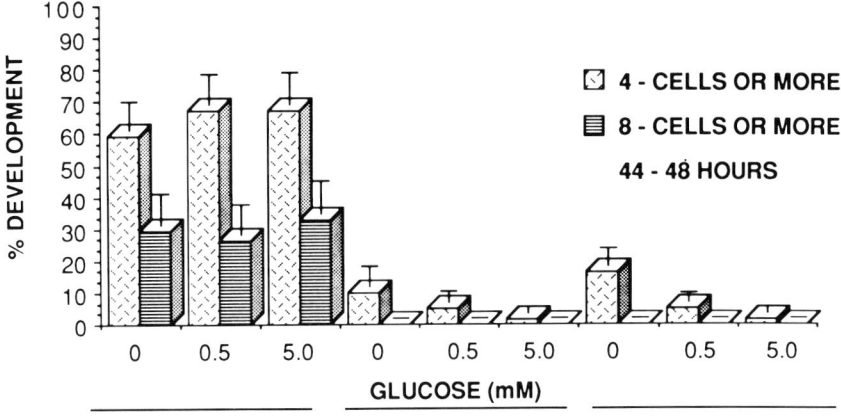

Figure 2. Effects of glucose and Pi concentrations on development of 2-cell hamster embryos in vitro. Results of a 3 X 3 factorial experiment with glucose at 0, 0.5 or 5 mM and Pi at 0, 0.1 or 0.35 mM. Culture medium was TLP-PVA with 20 amino acids. Values are means ± S.E.M. of 15 trials. Embryo development assessed after culture for 1 day (upper figure) and 2 days (lower figure). Data from (14).

Development of 2-cell hamster embryos was also tested in the presence of 0.1 mM deoxyglucose. The results we obtained are hard to explain, because deoxyglucose caused inhibition of development even in the absence of Pi; paradoxically, deoxyglucose relieved the inhibition caused by the presence of Pi (14). In this experiment, about 50% of the cultured 2-cell embryos developed as far as the morula stage (-Pi, ± glucose); however, no morulae formed in the presence of Pi. Although the Crabtree effect no longer seems to be consistent with the data on inhibition of hamster embryo development in vitro, we have been unable to find an alternative explanation for our results.

Effects of Phosphate on 4-Cell Embryos

Since our earlier experiments with 8-cell embryos (7) did not appear to show inhibitory effects of Pi or glucose (an erroneous conclusion: see below), whereas 2-cell embryos clearly were inhibited by these substances (14), we tested 4-cell hamster embryos for sensitivity to Pi. Hamsters were placed in a reverse daylight room so that 4-cell embryos could be collected at convenient times from superovulated mated females. We found that 4-cell hamster embryos, which block totally in standard culture media such as TLP-PVA with four amino acids, cleaved readily in this medium in the absence of glucose and Pi (21).

Several different experiments were done with 4-cell embryos to examine the characteristics of the Pi block on development. In one experiment, 4-cell embryos were randomly allocated to a control treatment (-Pi) and five experimental treatments containing increasing amounts of Pi from $1\mu M$ to $350\mu M$. After culture for 24 hours, about 50% of the 4-cell embryos developed to early or late blastocysts in the control medium (-Pi). In striking contrast, embryo development was suppressed in the presence of Pi, with no significant differences between any of the Pi concentrations; on average, only 11% or less of the embryos developed into blastocysts. It is remarkable that maximum inhibition of embryo development was observed at only 1μM Pi, and surprising that there was no dose response effect. The inhibitory effect of Pi was not due to some contaminant, e.g., heavy metal ions. Several different analytical grade sodium phosphate preparations all inhibited embryo development to the same degree; also, the levels of other ions present were found to be within the manufacturers' limits for contaminants. Furthermore, additional experiments showed that inclusion of EDTA in the culture medium did not overcome the inhibitory effect of Pi (21).

Another experiment showed that inhibition by Pi (in the absence of glucose) was largely confined to the 4-cell stage. Embryos collected at the mid 4-cell stage (46 hours post egg activation: 7) were cultured for varying lengths of time in glucose- and Pi-free TLP-PVA medium (with four amino acids) to allow development; then they were placed into medium containing 0.35 mM Pi (24 hours total incubation) to test their susceptibility to Pi inhibition (21). Four-cell embryos were completely inhibited by Pi, but mid and late 8-cell stages were able to develop into blastocysts in the presence of Pi. Early 8-cell embryos were intermediate between the 4-cell and mid 8-cell embryos, showing partial inhibition by Pi. Thus, hamster embryos escape from inhibition by Pi (in the absence of glucose; see below) after the third cleavage division. This experiment again illustrates how variable the response of hamster embryos can be, even within the same cleavage stage (e.g., 8-cell stage).

In a third experiment, we asked the question whether the inhibition was specific to Pi or if another divalent ion (SO_4) could achieve the same effect. This was not the case: when SO_4 was added at concentrations from 0.35 mM - 5.6 mM to medium without glucose and Pi, there were no significant differences compared to the high control (Pi- and SO_4-free medium) in terms of percent total blastocysts formed (34-44%). In contrast, the response in the low control containing 0.35mM Pi was significantly reduced (8% blastocysts). Thus, the inhibitory effect seems to be specific to Pi, but we still do not understand the mechanism of this inhibition.

Relationship of In Vitro Data to the Normal Milieu

How do these experimental data obtained with cultured 2- and 4-cell hamster embryos relate to the in vivo situation? It is possible that Pi is prevented from exerting an inhibitory effect on embryos by the presence of some other component of the oviductal environment. Alternatively, perhaps the Pi in oviduct secretions is not free to interact with embryos. Thus, the presence of Pi in the oviductal environment does not automatically mean that the in vitro inhibitory effects that we have described are artifacts of the culture environment, although this is certainly a possibility.

Transfer of Cultured Embryos to Recipients

It is important to transfer cultured embryos to recipient females in order to evaluate their viability. This has been done with hamster 8-cell embryos that were cultured from the 2-cell stage in our modified culture medium (TLP-PVA without Pi or glucose, with 20 amino acids; now called "HECM-1": 14). In a preliminary study using five pseudopregnant recipient female hamsters, 178 embryos were transferred after culture for approximately 24 hours (22). About half of the transferred embryos developed into fetuses (observed in two recipients on day 11 of pregnancy by laparotomy) and 23% of the cultured embryos survived to live pups. How much of the total losses are due to abnormal embryo development in vitro and how much are due to embryo transfer-related factors (including asynchrony) remains to be determined.

In two of the five cases, an albino female was mated to an albino vasectomized male to provide a genetic marker. Hamster embryos from golden coat color parents were cultured from 2-cells to about the 8-cell stage in HECM-1, then transferred to this albino recipient female; the 19 young produced had golden coats, confirming that they were derived from the cultured, transferred embryos.

Thus, some of the embryos that were able to develop through two cleavage divisions from the 2-cell stage in chemically-defined culture medium retained their viability. Under these conditions, neither exogenous protein, glucose or Pi was required for development.

EFFECTS OF GLUCOSE AND Pi ON 8-CELL EMBRYOS

In experiments similar to those described for 2- and 4-cell stages, effects of glucose and Pi were tested on development of 8-cell hamster embryos, using TLP-PVA medium containing Pi and only four amino acids (Phe, Ileu, Met, Gln). It was immediately apparent that glucose inhibited 8-cell hamster embryo development in vitro. In the absence of glucose, about 85% of 8-cell embryos developed into blastocysts but even 0.25 mM glucose depressed this to 55%, with no further inhibitory effect of glucose at least up to 1 mM (23).

In a series of factorial experiments, different combinations of energy substrates were used to examine the substrate preferences of cultured hamster 8-cell embryos. In modified medium (T-PVA) containing no energy substrates, i.e., no lactate, glucose, amino acids or pyruvate, 19% of 8-cell embryos were still able to reach the blastocyst stage (23). Remarkably, 5 mM glucose depressed the response to 7.9% ($p < 0.0005$). Thus, glucose is inhibitory to hamster 8-cell embryo development even in the absence of all other energy substrates. In the same experiment, inclusion of lactate (10 mM) as the sole energy substrate raised the proportion of blastocysts to 61.7% but again, addition of glucose severely depressed the response (18.8%, $p < 0.0005$). When lactate and glucose were used together, the percentage of embryos that developed into blastocysts (18.8%) was not different than in the complete absence of external energy substrates.

In a similar experiment (23), 23% of 8-cell embryos cultured in T-PVA without any energy substrates developed to the blastocyst stage, but when glucose was added as the sole external energy source, only 5% of embryos reached the blastocyst stage ($p < 0.0005$). Addition of the four standard amino acids (Phe, Ileu, Met, Gln) to the basic medium raised the percentage of blastocysts that formed to 60.5% but again, glucose depressed this response to 30.6% (p <0.0005). When glucose and amino acids were included together, the percentage of blastocysts that formed (30.6%) was not significantly different than when no energy substrates were present.

After a number of similar experiments, it became clear that glucose inhibits development of 8-cell hamster embryos both in the absence and in the presence of any of the other substrates tested (pyruvate, lactate or amino acids). However, these substrates ameliorate the inhibitory effects of glucose on embryo development (23). All of these experiments were done in the presence of Pi.

Comparison of Pi Effects on Different Stages of Hamster Embryo Development

The effect of Pi on 8-cell hamster embryo development was different than in the case of 2- and 4-cell embryos. Using medium HECM-2 (this is the more convenient name given to TLP-PVA without Pi and glucose), when either pyruvate or amino acids or lactate were included as the exogenous energy substrate, about 90% of cultured 8-cell embryos developed to the blastocyst stage (20). Addition of either Pi (0.35 mM) or glucose (5 mM) separately caused only a slight decrease in the percentage of blastocysts developing (to 70-80%). However, when Pi and glucose were added together, there was a dramatic drop in the percentage of blastocysts that developed from cultured 8-cell embryos. Only 50% of the embryos reached the blastocyst stage when pyruvate was the alternative energy substrate, and only 10-20% when either amino acids or lactate were used. Thus, at the 8-cell stage of development, Pi and glucose act synergistically to block embryo development (20, 23), in contrast to the 2- and 4-cell stages in which Pi is a powerful inhibitor of development in the absence of glucose, although glucose still inhibits only in the presence of Pi (14). This synergistic effect of pyruvate and glucose explains the low percentage of blastocysts developing from cultured 8-cell embryos in our earlier work using TALP medium (7).

AMINO ACID EFFECTS ON HAMSTER EMBRYO DEVELOPMENT

Pi and glucose are by no means the only components of culture media capable of inhibiting hamster embryo development. Although certain amino acids are clearly required for development of hamster embryos in vitro, other amino acids are inhibitory. Development of hamster 2-cell embryos into blastocysts within 48 hours was greater in HECM-1 medium containing 20 amino acids than in the HECM-2 variant containing only 4 amino acids (26% vs. 13%, respectively, p< 0.003: 24). However, when embryo culture was begun at the 8-cell stage, more late blastocysts formed in HECM-2 than in HECM-1 (77% vs. 39%, respectively, p< 0.003: 24). This discrepancy indicates that some of the amino acids in the total of 20 are inhibitory to embryo development. We are now conducting a series of experiments in which the stimulatory or inhibitory effects of single amino acids are being evaluated.

SUMMARY

Our results demonstrate that the hamster embryo behaves quite differently in culture than the classic model, the mouse embryo. Mouse embryos can develop from the 2-cell to the 8-cell stage in the presence of either pyruvate or lactate, and although these embryos cannot utilize glucose until the 8-cell stage, it is not inhibitory (8). From the 8-cell stage onwards, either pyruvate or lactate or glucose can be used to support development to the blastocyst stage. As a result, using conventional culture media containing all 3 substrates, development of mouse 2-cell embryos can progress to the blastocyst stage at a very high frequency (25). In marked contrast, our data show that cultured hamster embryos are inhibited by the presence of glucose at all stages of development from 2-cells to the blastocyst stage. However, this inhibition is dependent on the presence of Pi, which is a very powerful inhibitor of development of cultured hamster embryos and is the primary cause of the blocks to development in this species. At both the 2-cell and 4-cell stages, Pi inhibits embryo development independently of the presence of glucose; however, at the 8-cell stage, Pi (at 0.35 mM) is markedly inhibitory only when glucose is also present. We have no satisfactory explanations for these results as yet.

In addition to effects of Pi and glucose, we find that specific amino acids can have pronounced stimulatory or inhibitory effect on hamster embryo development in vitro. A considerable amount of work is needed to identify which of the amino acids are stimulatory and which are inhibitory, and whether or not the responses of embryos to particular amino acids change as development proceeds. This research would be helped considerably if we knew the amino acid composition of hamster oviduct fluid. Attempts are being made to obtain this information.

In conclusion, we are now able to obtain development of hamster embryos in vitro to a substantial degree, although complete preimplantation development (1-cell to blastocyst) in vitro is still not possible. At the present time, approximately 30% of hamster 2-cell embryos can reach the blastocyst stage in vitro when appropriate culture conditions are provided, including the absence of Pi and glucose, the presence of 20 amino acids and culture under 10% CO_2 (26, 27). However, in vitro fertilized 1-cell embryos still cleave only once, forming abnormal looking 2-cell embryos with somewhat swollen blastomeres, and none of these embryos develop to the 4-cell stage.

We still do not know the reason for the failure of in vitro fertilized 1-cell hamster embryos to develop in culture. It is possible that zygotes produced by this means are abnormal as a result of fertilization occurring in vitro, or possibly our culture medium is still inadequate for development of very early 1-cell embryos. Alternatively, perhaps a combination of these two explanations is the answer. Work is ongoing in our laboratory to resolve these questions. When we are successful in supporting development of in vitro fertilized hamster embryos to the blastocyst stage in vitro, a wide range of basic studies can be undertaken, including analyses of the developmental consequences of in vitro fertilization in this species, and of the possible contribution of delayed fertilization to developmental anomalies.

ACKNOWLEDGMENTS

I would like to thank the individuals in my laboratory at the University of Wisconsin-Madison whose hard work and perseverance underlies the research described in this chapter: Ed Carney, Pat Farrell, Susan Harmon-McKiernan, Lorraine Leibfried-Rutledge, Helena Monis, Polani Seshagiri and Scott Schini. I thank Rick Gneiser for the artwork and photography.

REFERENCES

1. Bavister BD (1988). Role of oviductal secretions in embryonic growth in vivo and in vitro. Theriogenology 29:143.
2. Yanagimachi R, Chang MC (1963). Fertilization of hamster eggs in vitro. Nature (London) 200:281.
3. Yanagimachi R, Chang MC (1964). In vitro fertilization of golden hamster ova. J Exp Zool 156:361.
4. Whittingham DG, Bavister BD (1974). Development of hamster eggs fertilized in vitro and in vivo. J Reprod Fert 38:489.
5. Sato A, Yanagimachi R (1972). Transplantation of preimplantation hamster embryos. J Reprod Fertil 30:329.
6. Hoppe RW, Bavister BD (1983). Evaluation of the fluorescein diacetate (FDA) vital dye viability test with hamster and bovine embryos. Anim Reprod Sci 6:323.
7. Bavister BD, Leibfried ML, Lieberman G (1983). Development of preimplantation embryos of the golden hamster in a defined culture medium. Biol Reprod 28:235.
8. Biggers JD (1987). Pioneering mammalian embryo culture. In Bavister BD (ed.): "The Mammalian Preimplantation Embryo," New York: Plenum Press, p 1.

9. Gwatkin RB, Haidri AA (1973). Requirements for the maturation of hamster oocytes in vitro. Exp Cell Res 76:1.
10. Farrell PS, Bavister BD (1984). Short-term exposure of 2-cell hamster embryos to collection media is detrimental to viability. Biol Reprod 31:109.
11. Bavister BD (1987). Studies on the developmental blocks in cultured hamster embryos. In Bavister BD (ed.): "The Mammalian Preimplantation Embryo," New York: Plenum Press, p 219.
12. Bavister BD (1989). A consistently successful procedure for in vitro fertilization of golden hamster eggs. Gamete Res 23: (In Press).
13. Schini SA, Bavister BD (1988). Development of golden hamster embryos through the two-cell block in chemically defined medium. J Exp Zool 245:111.
14. Schini SA, Bavister (1988). Two-cell block to development of cultured hamster embryos is caused by phosphate and glucose. Biol Reprod 39:1183.
15. Kaye PL (1986). Metabolic aspects of the physiology of the preimplantation embryo. In Rossant J, Pedersen RA (eds.): "Experimental Approaches to Mammalian Embryonic Development," Cambridge, UK: Cambridge University Press, p 267.
16. Rogers BJ, Yanagimachi R (1975). Retardation of guinea pig sperm acrosome reaction by glucose: the possible importance of pyruvate and lactate metabolism in capacitation and the acrosome reaction. Biol Reprod 13:568.
17. Parrish JJ, Susko-Parrish JL, First NL (1986). Capacitation of bovine sperm by oviduct fluid or heparin is inhibited by glucose. J Androl 7:22P.
18. Crabtree HG (1929). Observations on the carbohydrate metabolism of tumours. Biochem J 23:536.
19. Koobs DH (1972). Phosphate mediation of the Crabtree and Pasteur effects. Science 178:127.
20. Seshagiri PB, Bavister BD (1989). Phosphate is required for inhibition by glucose of dev-

elopment of hamster 8-cell embryos in vitro. Biol Reprod (In Press).
21. Monis H, Bavister BD (in preparation).
22. Schini SA, Bavister BD (in preparation).
23. Seshagiri PB, Bavister BD (1989). Glucose inhibits development of hamster 8-cell embryos in vitro. Biol Reprod (In Press).
24. Seshagiri PB, Bavister BD. Relative developmental abilities of hamster 2- and 8-cell embryos cultured in HECM-1 and HECM-2. J Reprod Fertil (submitted).
25. Biggers JD, Whitten WK, Whittingham DG (1971). The culture of mouse embryos in vitro. In Daniel JC, Jr (ed.): "Methods in Mammalian Embryology," San Francisco: WH Freeman, p 86.
26. Carney EW, Bavister BD (1987). Regulation of hamster embryo development in vitro by carbon dioxide. Biol Reprod 36:1155.
27. Harmon-McKiernan S, Bavister BD (in preparation).

GENE EXPRESSION REQUIRED FOR BLASTOCOEL FORMATION IN THE MOUSE[1]

Gerald M. Kidder and Andrew J. Watson

Department of Zoology, The University of Western Ontario, London, Ontario, Canada N6A 5B7

ABSTRACT Blastocoel formation in the mouse involves active fluid transport across the trophectoderm layer, a process which is sensitive to ouabain, a specific inhibitor of Na^+,K^+-ATPase (the sodium pump). This and other evidence marks Na^+,K^+-ATPase as a critical agent in cavitation. We have used antibody and cDNA probes for the catalytic subunit of this enzyme to test the hypothesis that expression of the genes encoding the sodium pump is an important determinant of the timing of cavitation. We find that the catalytic subunit first becomes detectable in the late morula, just prior to the onset of blastocoelic fluid accumulation. It assumes a juxtacoelic distribution in the early blastocyst where it is concentrated in the basolateral domain of the mural trophectoderm, including its projections covering the inner cell mass. In contrast, the mRNA encoding the catalytic subunit is detectable as early as the 8-cell stage, and persists into the early blastocyst. The expression of this gene is thus both an indicator of trophectoderm differentiation and a likely determinant of cavitation, although the timing of the enzyme's appearance appears to involve post-transcriptional regulatory processes.

---

[1]This work was supported by grants from NSERC Canada and the National Institutes of Health.

INTRODUCTION

In the mouse, fluid accumulation to form the blastocoel is known to depend on the development of two critical plasma membrane functions: intercellular communication via membrane channels (gap junctions), and trans-trophectodermal sodium transport driven by Na^+,K^+-ATPase (the sodium pump). Embryos defective in gap junctional communication are unable to maintain compaction and subsequently fail to cavitate (1, 2). Embryos treated with ouabain, an inhibitor of Na^+,K^+-ATPase, are unable to accumulate blastocoelic fluid (3-5). We have begun to analyze the expression of the genes encoding these two functions using antibodies and cDNA probes. In the present communication we report recent findings concerning the expression of the gene encoding the catalytic subunit of Na^+,K^+-ATPase.

Cavitation is sensitive to agents which disrupt transcription or protein synthesis, demonstrating that, like most other events of preimplantation development, it requires new gene products at least some of which must be encoded by the embryonic genome. The transcriptional events necessary for initiating blastocoel formation are apparently complete about 5-7 hours before cavitation begins, since after this time transcriptional blockade fails to delay the onset of fluid accumulation (6). This finding implies that some of the agents active in cavitation are products of relatively stable mRNAs which accumulated earlier, a suggestion consistent with the fact that the average half life of mRNA increases after compaction (7). From what we know thus far, then, it is likely that the timing of morphogenetic events like cavitation is dependent at least in part on post-transcriptional regulatory processes (6).

In focusing our attention on specific genes involved in blastocoel formation, we hope to provide a more accurate picture of the programming of cavitation. The genes encoding the subunits of Na^+,K^+-ATPase are an obvious choice: in addition to the evidence of ouabain sensitivity mentioned above, the juxtacoelic distribution of the catalytic subunit in expanding blastocysts further substantiates the role of this enzyme in transtrophectodermal sodium and fluid transport (8). We have now confirmed that this subunit is largely restricted to the mural trophectoderm, and that its mRNA is present well in advance of the time when the protein itself first becomes detectable.

METHODS

Our procedure for detection of the Na^+, K^+-ATPase catalytic subunit by immunofluorescence has been described previously (8). As in our earlier work, we used a polyclonal antibody raised against the catalytic subunit purified from rat kidney, supplied by Dr. Mark Ellisman (University of California, San Diego).

Transmission electron micrographs were obtained from thin sections of early blastocysts which had been fixed in 1.5% glutaraldehyde and 0.5% paraformaldehyde, and embedded in Spurr's resin as described previously (9).

Total RNA was extracted from batches of embryos that had been washed free of collecting medium using PBS-PVP (phosphate-buffered saline containing 3 mg/ml polyvinylpyrollidone), quick-frozen on dry ice in tiny volumes (less than 10 microliters) of PBS-PVP, and stored at -70°C. We used the single-step RNA extraction method of Chomczynski and Sacchi (10) which we modified slightly by reducing the extraction volume ten-fold and by eliminating additional alcohol precipitation steps. The RNA pellets (along with 20 micrograms of carrier tRNA) were denatured by heating to 55°C in 50% formamide, 6.5% formaldehyde, and then separated by electrophoresis in 1% agarose gels containing 660 mM formaldehyde. The separated RNAs were transferred to NitroPlus 2000 membrane (Micron Separations Inc., Westborough, MA) by capillary blotting. The blots were probed with cDNAs labeled with ^{32}P using the random primers method; hybridization was at 42°C for 40-44 hours in 5X SSPE, 50% formamide, 5X Denhardt's, 10% dextran sulfate, and 0.1 mg/ml denatured salmon sperm DNA (11). After hybridization the blots were washed to moderate stringency (0.5X SSC, 0.5% SDS, 65°C). The sodium pump catalytic subunit cDNA (rb5) was supplied by Dr. Robert Levenson (Yale University); it was obtained by screening a rat brain cDNA library with antiserum raised against Na^+, K^+-ATPase from rat kidney (12). The histone H3.2 cDNA was supplied by Dr. Gilbert A. Schultz (University of Calgary). An LKB UltroScan XL laser densitometer was used to quantify the hybridization signals.

RESULTS

Using indirect immunofluorescence, we have not been able to detect the catalytic subunit of Na^+, K^+-ATPase

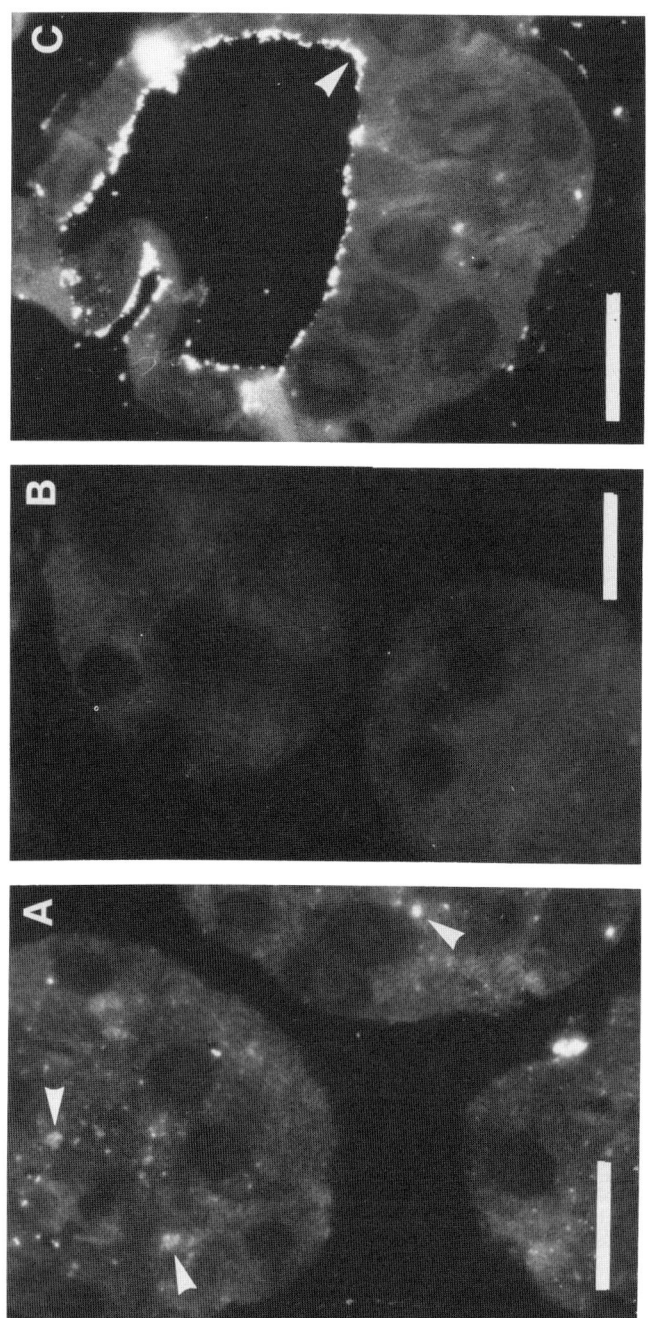

FIGURE 1. The catalytic subunit of Na^+,K^+-ATPase first becomes detectable by immunofluorescence in late morulae. Embryos collected 84 hours post-hCG were embedded in polyethylene glycol and sectioned at a thickness of 0.5 microns. Sections of late morulae are shown in A and B; an early blastocyst section is shown in C. A few of the more prominent sites of immunoreactivity are indicated by arrowheads. Each scale bar equals 20 microns.

Gene Expression and Blastocoel Formation 101

prior to the late morula stage. As shown in Figure 1A, late morulae (84 hours post-hCG) lacking a nascent blastocoel cavity have immunoreactivity in discrete intracellular foci. It is common, however, for a population of late morulae such as this to include some embryos which do not show positive immunoreactivity (Figure 1B). Sections treated with nonimmune serum, or with secondary antibody only, were likewise negative (8). With the first appearance of an extracellular fluid-filled cavity (Figure 1C), the catalytic subunit assumes a juxtacoelic distribution, i.e., it lines the cavity on the inner (basolateral)

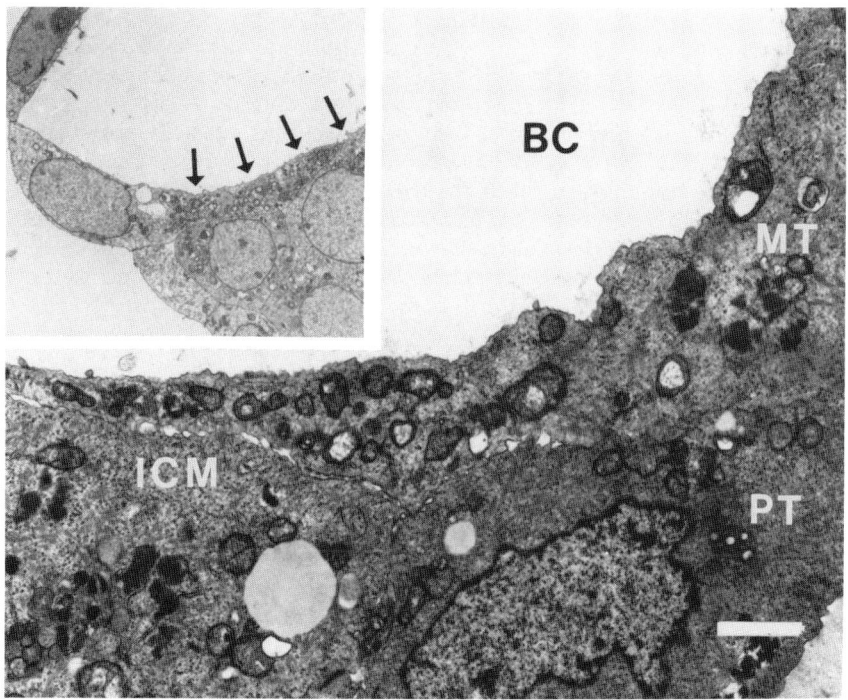

FIGURE 2. Early blastocysts with a juxtacoelic distribution of catalytic subunit immunoreactivity have projections of the mural trophectoderm covering the ICM adjacent to the blasocoel. MT, mural trophectoderm; PT, polar trophectoderm; ICM, inner cell mass; BC, blastocoel cavity; the scale bar equals 2 microns. In the inset a trophectodermal projection is indicated by arrows.

surface of the mural trophectoderm.

In early blastocysts the catalytic subunit immunoreactivity appears to cover the surface of the ICM adjacent to the blastocoel, and this immunoreactivity is often continuous with that in the mural trophectoderm (Figure 1C, arrowhead). It has been reported that the surface of the ICM is overlain at this stage by projections of the mural trophectoderm (13), and we suggested that such projections, rather than the ICM itself, could be providing the immunoreactive sites (8). We have now verified this by electron microscopy. Early blastocysts having the characteristic juxtacoelic distribution of the enzyme do have projections of the mural trophectoderm covering the surface of the ICM (Figure 2). This finding supports our contention that abundant expression of Na^+,K^+-ATPase is a characteristic feature of mural trophectoderm, as distinct from polar trophectoderm and ICM.

To search for the mRNA encoding the catalytic subunit we prepared total RNA from batches of 1,000 embryos from various stages for northern blot analysis. In the mouse there are three known genes, residing on separate chromosomes, which encode isoforms of the catalytic subunit (14); each of these genes has a particular pattern of expression in adult organs (15). We probed northern blots with a partial cDNA (rb5) corresponding to the $\alpha 1$ subunit, which is expressed in most (if not all) tissues (15). This 1.2 kb cDNA (Figure 3) covers a region of the mRNA which is highly conserved among the three isoforms since it encodes both the phosphorylation and ATP binding sites; it would thus be expected to hybridize with any of the

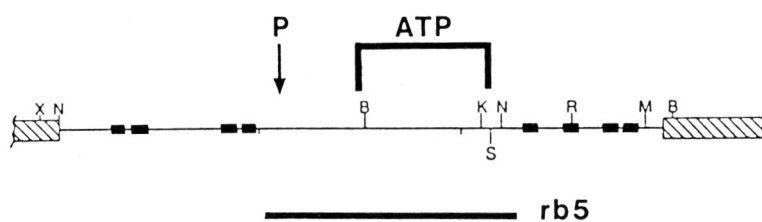

FIGURE 3. The cDNA rb5 covers a portion of the $\alpha 1$ catalytic subunit mRNA which encodes the phosphorylation site (P) and the region making up the ATP binding site. The other letters indicate restriction enzyme sites in the cDNA and the solid blocks represent sequences which encode putative membrane spanning domains.

three catalytic subunit mRNAS at moderate stringency. In order to provide an internal standard for RNA recovery and integrity, we hybridized the blots simultaneously with a histone H3.2 cDNA which we had shown previously recognizes a single RNA (0.6-0.7 kb) in all preimplantation stages. As shown in Figure 4A, the rb5 cDNA hybridized to

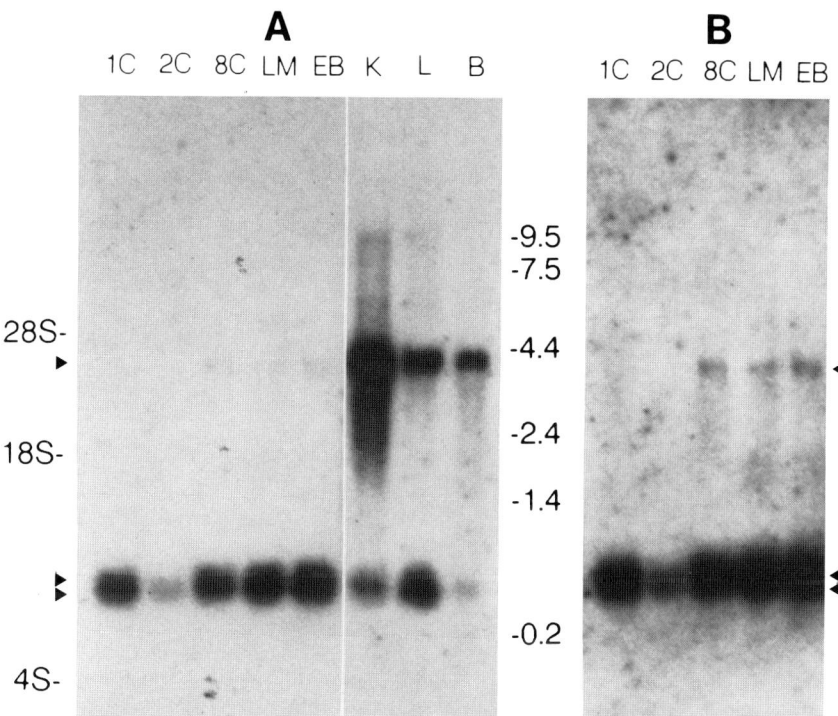

FIGURE 4. The mRNA encoding the catalytic subunit can be detected as early as the compacted 8-cell stage. Each embryo lane contained total RNA from 1,000 embryos; the kidney (K), liver (L), and brain (B) lanes contained 5-10 micrograms of total RNA. The developmental times were: 1C, zygotes (26 hr post-hCG); 2C, 2-cell (39 hr); 8C, 8(+)-cell (76 hr); LM, late morula (86-92 hr); EB, early blastocyst (88-94 hr). The autoradiogram was exposed first for three days (A), then for two weeks (B). The catalytic subunit cDNA detected a single 4 kb RNA (arrowhead); the histone mRNA band is indicated by double arrowheads.

a single 4 kb RNA in 8-cell and later stages that is the same size as the corresponding mRNA in kidney, liver, and brain. No hybridization signal was detected in zygotes or 2-cell embryos even after two weeks' exposure of the autoradiogram (Figure 4B). The histone cDNA, on the other hand, detected its cognate mRNA producing a strong signal in the zygote sample; the amount of this message declines in the 2-cell stage to accumulate again in later stages as was reported previously (16). Comparison of the developmental profiles of these two mRNAs indicates that mRNA for the catalytic subunit of Na^+,K^+-ATPase is relatively rare in zygotes and 2-cell embryos, representing the interval between 26 and 39 hours post-hCG, and accumulates after activation of embryonic transcription in the 2-cell stage. Densitometric scans revealed that by the compacted 8-cell stage (76 hours post-hCG), the catalytic subunit mRNA had accumulated to 73% of its level in early blastocysts, whereas histone mRNA had reached only 50% of its blastocyst level.

DISCUSSION

The immunofluorescence results reported here indicate that the catalytic subunit of Na^+,K^+-ATPase becomes detectably abundant in the late morula, just prior to the onset of cavitation. Earlier stages (i.e., 4- and 8-cell embryos, ref. 8) are negative for immunoreactivity with this antiserum, as are some late morulae obtained in an 84 hour post-hCG flush. The fact that an 84 hour flush contains a mixture of positive and negative late morulae implies that the epitopes recognized by this antiserum become abundant or are unmasked during a narrow window of time. Such heterogeneity was not seen among the early blastocysts of the same flushes. The simplest interpretation of these findings is that the catalytic subunit accumulates rapidly in the few hours before the onset of extracellular fluid accumulation, appearing first in a transient intracellular location, then being inserted into the basolateral plasma membrane of the mural trophectoderm. Projections of this membrane covering the surface of the ICM complete the extension of sodium pump-containing membrane to form a complete lining around the expanding blastocoel. This enzyme is thus a characteristic feature of mural trophectoderm in the blastocyst, and the timing of expression of the genes encoding its subunits is likely to be one important determinant of the timing of cavitation. One prediction of this model is that the rate of synthesis of the catalytic subunit should increase markedly in late morulae and during blastocyst expansion, as has been demonstrated in the rabbit (17).

The mRNA for the catalytic subunit, on the other hand, can be detected as early as the compacted 8-cell stage, 76 hours post-hCG. Our failure to detect it in 26 hour zygotes and in the 2-cell stage probably reflects the insensitivity of the northern blot assay, since Na^+,K^+-ATPase is ubiquitous in mammalian cells and all developmental stages would be expected to maintain at least a low level of expression of the genes encoding its subunits. We suggested in our earlier report (8) that blastocoel formation requires a much greater abundance of sodium pumps, which is brought about in the hours leading up to cavitation. The northern blot data demonstrate that the catalytic subunit mRNA does accumulate during the cleavage stages, confirming that the increase in abundance of the enzyme later in development involves transcription of the embryonic genome. It is interesting to note that by 76 hours post-hCG this mRNA has accumulated to more than 70% of its level in the early blastocyst, which can explain the relative insensitivity of cavitation to transcriptional blockade after this time (6). Despite this, the catalytic subunit itself cannot be detected by our immunofluorescence procedure until around 84 hours post-hCG. It may be that the principal determinant of the timing of blastocoel formation is not the transcription of Na^+,K^+-ATPase genes, but the activation of translation of an already abundant pool of their transcripts.

ACKNOWLEDGEMENTS

We would like to thank Cindy Pape for her expert technical assistance, Drs. Mark Ellisman, Robert Levenson, and Gilbert Schultz for supplying antibodies or cDNAs, Gunnar Valdimarsson and Paul DeSousa for useful discussions concerning interpretation of our findings, and Lori Dos Santos for typing the manuscript.

REFERENCES

1. Lee S, Gilula, NB, Warner, AE (1987). Gap junctional communication and compaction during preimplantation stages of mouse development. Cell 51:851.
2. Buehr M, Lee S, McLaren A, Warner A (1987). Reduced gap junctional communication is associated with the lethal condition characteristic of DDK mouse eggs fertilized by foreign sperm. Devel 101:449.
3. DiZio SM, Tasca RJ (1977). Sodium-dependent amino acid transport in preimplantation mouse embryos. Dev Biol 59:198.
4. Wiley LM (1984). Cavitation in the mouse preimplantation embryo: Na/K-ATPase and the origin of nascent blastocoele fluid. Dev Biol 105:330.
5. Manejwala FM, Cragoe EJ Jr, Schultz RM (1989). Blastocoel expansion in the preimplantation mouse embryo: role of extracellular sodium and chloride and possible apical routes of their entry. Dev Biol (in press).
6. Kidder GM, McLachlin JR (1985). Timing of transcription and protein synthesis underlying morphogenesis in preimplantation mouse embryos. Dev Biol 112:265.
7. Kidder GM, Pedersen RA (1982). Turnover of embryonic messenger RNA in preimplantation mouse embryos. J Embryol Exp Morph 67:37.
8. Watson AJ, Kidder GM (1988). Immunofluorescence assessment of the timing of appearance and cellular distribution of Na/K-ATPase during mouse embryogenesis. Dev Biol 126:80.
9. Kidder GM, Barron DJ, Olmsted JB (1988). Contribution of midbody channels to embryogenesis in the mouse. Analysis by immunofluorescence. Roux's Arch Dev Biol 197:110.
10. Chomczynski P, Sacchi N (1987) Single-step method of RNA isolation by acid guanidinium thiocyanatephenol-chloroform extraction. Anal Biochem 162:156.
11. Maniatis T, Fritsch EF, Sambrook J (1982). "Molecular Cloning." Cold Spring Harbor, NY: Cold Spring Harbor Laboratory.
12. Schneider JW, Mercer RW, Caplan M, Emanuel JR, Sweadner KJ, Benz EJ Jr, Levenson R (1985). Molecular cloning of rat brain Na,K-ATPase α-subunit cDNA. Proc Natl Acad Sci USA 82:6357.

13. Fleming TP, Warren PD, Chisholm JC, Johnson MH (1984). Trophectodermal processes regulate the expression of totipotency within the inner cell mass of the mouse expanding blastocyst. J Embryol Exp Morph 84:63.
14. Kent RB, Fallows DA, Geissler E, Glaser T, Emanuel JR, Lalley PA, Levenson R, Housman DE (1987). Genes encoding α and β subunits of Na,K-ATPase are located on three different chromosomes in the mouse. Proc Natl Acad Sci USA 84:5369.
15. Herrera VLM, Emanuel JR, Ruiz-Opazo N, Levenson R, Nadal-Ginard B (1987). Three differentially expressed Na,K-ATPase α subunit isoforms: structural and functional implications. J Cell Biol 105:1855.
16. Graves RA, Marzluff WF, Giebelhaus DH, Schultz GA (1985). Quantitative and qualitative changes in histone gene expression during early mouse embryo development. Proc Natl Acad Sci USA 82:5685.
17. Overstrom EW, Benos DJ, Biggers JD (1989). Synthesis of Na^+/K^+ATPase by the preimplantation rabbit blastocyst. J Reprod Fert 85:283.

THE ROLE OF INSULIN IN PREIMPLANTATION MOUSE DEVELOPMENT[1]

L.V. Rao, M. Farber, [*]R.M. Smith and S. Heyner

Department of Obstetrics and Gynecology
Albert Einstein Medical Center
Philadelphia, PA 19141 and

[*]Department of Pathology and Laboratory Medicine
University of Pennsylvania
Philadelphia PA 19104

ABSTRACT The critical elements of cellular proliferation and differentiation in mammalian embryogenesis may be influenced by hormones and growth factors from either embryonic or maternal sources. Insulin is one of the most important hormones that influences fetal growth and metabolism; however, little is known concerning its interaction with cells of the preimplantation embryo. We report here that incubation of preimplantation mouse embryos in physiological levels of insulin resulted in an increase in cell number, and stimulated the incorporation of labelled precursors into nucleic acids. Colloidal gold-labelled insulin was used in conjunction with high-resolution electron microscopy allowed the visualization of insulin binding and internalization, and permitted the elucidation of the endocytotic pathway. Immunocytochemical studies using gold-labelled anti-insulin antibodies revealed that the source of insulin during early development is maternal. These results document that insulin exerts a mitogenic effect during preimplantation development in the mouse.

---

[1]These studies were supported by NIH grants HD 23511 and DK 19525.

INTRODUCTION

Early embryo development is regulated primarily by the expression of specific genetic programs within the cells, but also requires a continuous supply of energy, hormones and growth factors. There is evidence that early development is dependent in part on specific environmental conditions provided by the maternal genital tract, at least for energy substrates and amino acids [1]. There is little information concerning the role of hormones and growth factors in preimplantation stages of development. Among peptide growth factors, insulin and its receptor have been detected early in development in a number of non-mammalian species [reviewed in 2].

Insulin is a small protein with a molecular weight of approximately 6000. It contains 51 amino acids, arranged in two chains, A and B, linked by disulfide bridges. Amino acid sequences of insulin from a number of vertebrates show that the peptide is highly conserved in evolution. Insulin triggers a complex array of metabolic processes in a variety of mammalian cell types. It plays a key role in the intermediary metabolism of muscle and adipose tissue, and has important effects in the liver. In addition to its metabolic effects, there is evidence that insulin may regulate gene action. Insulin is capable of stimulating DNA synthesis and cell growth in a number of cell lines in vitro. There is evidence that insulin may influence preimplantation embryonic growth; it has been shown to stimulate protein synthesis in compacted mouse embryos [3], and to influence glucose transport in mouse embryos [4,5]. Using autoradiographic methods, Mattson et al. [6] were able to detect the expression of specific receptors that bind insulin and insulin-like growth factors (IGFs) as early as the morula stage of development. These results were confirmed at the mRNA level, using an mRNA phenotyping technique, which showed that transcripts encoding the insulin receptor became detectable between the 8-cell and morula stage in mouse embryos [7]. The presence of insulin receptors in early mouse embryos may be indicative of a developmental function of this peptide. In the present study, we report the effect of insulin on cell growth and nucleic acid synthesis at different stages of preimplantation mouse development. Colloidal gold-labelled insulin in conjunction with high resolution electron microscopy was used to study the uptake and distribution of insulin in different cellular compartments in preimplantation embryos. These studies showed that insulin is internalized by means of receptor-mediated endocytosis exclusively through coated pits.

METHODS

Animals and Embryo Recovery

Random-bred CD-1 mice (Charles River Breeding Laboratories) were used. Females aged 6-8 weeks of age were induced to superovulate and mated using standard methods [8]. Mating was verified by the presence of a vaginal copulatory plug the next morning (day 1 of pregnancy). Embryos of different developmental stages were flushed from the reproductive tract using M2 medium [9]. Embryos were cultured for at least 1 hour in medium M16 [9] at 37°C in an atmosphere of 5% CO_2 in humidified air. Culture media M2 and M16 contained bovine serum albumin (BSA) that was insulin-free.

Uptake and Incorporation of ^{3}H-Thymidine and ^{3}H-Uridine

Groups of 15-20 embryos (2-, 8-cell, morula and blastocyst) were flushed from the reproductive tract, and cultured in medium M16 containing 0.4% insulin-free BSA, in an atmosphere of 5% CO_2 in humidified air, at 37°C for 1 hour. They were then incubated for 1 hour in M16 in the presence or absence of insulin (4 ng/ml). Embryos were then transferred to M16 containing 50 uçi/ml ^{3}H-thymidine (methyl^{3}H-thymidine, Amersham) or ^{3}H-uridine (5,6^{3}H-uridine, Amersham) with or without insulin, at a concentration of 4 ng/ml, and incubated for a further 2 hours. Embryos were washed extensively with ice-cold medium M2 containing 1000x excess unlabelled thymidine or uridine, respectively, and placed in 25 ul of medium containing 2 mg/ml BSA. Following the addition of 25 ul of 10% trichloracetic acid (TCA), they were incubated on ice for 15-20 min and spun in a microfuge for 5 min at 4°C. The pellet was washed twice with 100 ul of cold 10% TCA, and the supernatants combined. TCA soluble radioactivity was determined by adding 10 ml Hydrofluor, and counting in an LKB liquid scintillation counter. The TCA insoluble pellet was dissolved in 30 ul 1N NaOH, acidified with 100 ul 1N HCl, and counted after the addition of 10 ml Hydrofluor.

Determination of Embryo Cell Number

Embryos at the 8-cell stage were flushed from the tract in M2 medium, and cultured for 48 hours in M16 containing 0 or 40 ng/ml insulin, with one change of medium after 24 hours. They were then scored under the dissecting microscope, and their developmental stage assessed. Embryos were mounted on glass slides and stained with Hoechst 33342 stain, using the procedure described by Pursel et al [10]. Briefly, embryos were placed on a glass slide, stained with trypan blue, and then with Hoechst 33342, and mounted in Flo-tex mounting medium. The nuclei of the embryos fluoresced brightly when examined by epi-fluorescence microscopy. The embryos were scored for the number of nuclei per embryo, and the number of cells was assumed to be equivalent to the number of nuclei.

Binding and Uptake of Gold-labelled Insulin

Mouse embryo developmental stages (2-, 8-cell, morula and blastocyst) were flushed from the reproductive tract, and incubated for 1 hour in M16 containing insulin-free BSA. They were then incubated for 45 min at 37°C in medium M16 (pH 7.4) containing gold-labelled insulin [11] equivalent to 200 ng/ml of native hormone. Specificity of binding was established by incubation of controls with an identical concentration of gold-labelled BSA, or in medium containing 40 ug/ml unlabelled insulin in addition to the gold-labelled ligand. Following incubation, embryos were washed extensively with ice-cold phosphate buffered saline containing 1% insulin-free BSA, and processed for light and electron microscopy. Thick sections (1 micron) were stained with toluidine blue (0.5% in 0.5% $Na_2B_4O_7$, pH 11) in order to assess developmental stage, and confirm orientation of the embryos. Sections were cut on an LKB ultramicrotome, stained alcoholic uranyl acetate and bismuth subnitrate, and examined in a JEOL 100CX electron microscope. Gold particle counts were made from thin sections cut across the embryos, and the distribution and location of the gold particles with respect to cellular compartments was evaluated.

Immunocytochemical Localization of Insulin

Embryos at the blastocyst stage were flushed from the reproductive tract, and either fixed immediately in buffered picric acid (BGPA) fixative, or cultured, either the presence of 200 ng/ml unlabelled insulin for 2 hours, or in the absence of insulin overnight. Following culture, the embryos were fixed in BGPA. After fixation, the embryos were washed extensively with phosphate buffered saline, and partially dehydrated with 70% ethanol. They were embedded in LR White resin (London Resin Co. Ltd.), and following polymerization, thick and thin sections were cut. Thick section were examined for purposes of embryo orientation and staging. Thin sections were mounted on naked nickel grids. The thin sections were incubated with guinea pig anti-porcine insulin IgG (Miles Laboratories), labelled with colloidal gold as described by Horisberger and Vauthey [12], and incubated at $4^\circ C$ overnight. Sections were then washed extensively with phosphate buffer and deionized water, stained with aqueous uranyl acetate, and analyzed as described above. In order to examine the reproductive tract for the presence of insulin, the contralateral uterine horn and oviduct of a female that had blastocysts flushed from one side was fixed in BGPA, and processed for immunocytochemical analysis as described for the blastocysts.

Specificity of the binding of the gold-labelled antibodies was established by testing the procedure in tissues known to have the antigen, and negative tissues.

RESULTS AND DISCUSSION

Previous studies utilizing immunofluorescence [13] and autoradiography [6] have shown that mouse preimplantation embryos express insulin receptors in a developmentally-regulated manner, beginning at the late 8-cell stage and increasing through the blastocyst. The present report confirms these results, and provides evidence that insulin exerts a metabolic effect at early stages in mammalian development. In addition, we show that insulin bound by the embryo originates in the maternal reproductive tract, in contrast to autocrine production of this peptide. The effect of insulin on the synthesis of DNA and RNA in embryos is shown in Fig. 1.

Incubation of embryos in insulin at physiological levels had no influence on total accumulation of precursors, showing that insulin did not affect transport into intact embryos. However, there was a significant increase in the incorporation into the TCA insoluble fraction, showing that the effect was specific for stimulation of DNA and RNA synthesis. The treatment was significant at the morula and blastocyst stages; embryos at the morula and blastocyst stages had approximately an 8% and 27% increase respectively, in incorporation of ^{3}H-thymidine into the TCA-insoluble fraction, while the corresponding results for the incorporation of ^{3}H-uridine were 54% increase at the morula stage, and 110% increase at the blastocyst stage.

In agreement with previous studies, the effect was variable at the 8-cell stage, and was not detectable at the 2-cell stage. The absence of a response at stages earlier than eight cells is consonant with a lack of receptor expression. These results provide evidence that insulin has a metabolic effect at in the earliest stages of mammalian development.

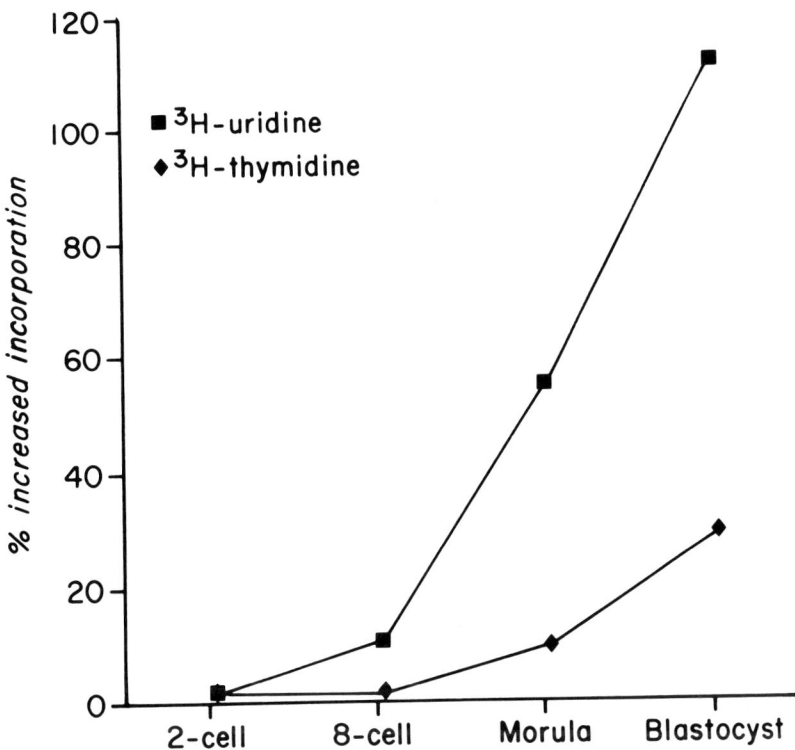

FIGURE 1. Effect of insulin on the mean percentage increase incorporation of ^{3}H-thymidine and ^{3}H-uridine into the acid-insoluble fraction of mouse preimplantation embryos.

The influence of insulin on cell proliferation and growth promotion have been described for a number of cell lines in vitro [14]. When the effect of culturing embryos for 48 hours in physiological concentrations of insulin was examined, a mitogenic effect of the hormone was apparent [Table 1].

TABLE 1
INSULIN EFFECT ON EMBRYO CELL NUMBER[a]

Control	Insulin 40 ng/ml	% Increase
40.8 ± 1.7 (n = 27)	52.8 ± 2.5 (n = 28)	29.3

[a]The effect of insulin (40 ng/ml) on the cell number of preimplantation mouse embryos grown for 48 hours in vitro. Values represent the mean number of nuclei per embryo ± the standard error of the mean.

The earliest event in insulin action on target cells involves binding to its receptor, followed by the activation of the tyrosine kinase activity associated with it [15]. The insulin receptor is present in virtually all mammalian tissues, although receptor density varies form as few as 40 receptors per cell on circulating erythrocytes, to more than 200,000 on adipocytes and hepatocytes [16]. In common with membrane receptors for steroid hormones, the insulin receptor has at least two functions. The first is to recognize the hormone, which is accomplished by binding the ligand with high affinity and a high degree of specificity. The second function is to produce a transmembrane signal that alters intracellular metabolism, and mediates the action of the hormone. Insulin receptors show great structural and functional homology in different tissues. The insulin receptor is a glycoprotein consisting of two 130,000 M_r subunits (alpha chain) and two 95,000 M_r subunits (beta subunits) in a sulfide-linked, transmembrane complex whose complete amino acid sequence has been established [17,18]. The structure of the two subunits is highly conserved. The alpha chain, which comprises the external domain of the receptor, binds insulin, and the cytoplasmic domain of the beta chain possesses tyrosine kinase activity, which results in autophosphorylation on tyrosine residues within seconds of ligand binding [19].

Conventional biochemical techniques have been of little value for the study of insulin receptors in preimplantation development, principally because of the small amount of material that is available. However, the use of specific antibodies in conjunction with immunofluorescent techniques allows the investigator to examine the expression of molecules of interest on very small tissue samples. This approach has been used successfully in a number of studies on mouse embryos. Antibodies directed against insulin were used in conjunction with indirect immunofluorescence to examine a developmental series of mouse embryos for the ability to bind insulin. Insulin binding could be detected first at the morula stage of development, and increased through the blastocyst stage [13]. Despite the developmentally-regulated appearance of the ability to bind insulin, the investigators had to exclude the possibility that the hormone was being bound non-specifically, by the extensive endocytotic system that develops around the eight-cell stage in the mouse embryo [20]. This possibility was excluded by the use of light microscopic autoradiography. These experiments showed that binding of ^{125}I-insulin occurred first at the morula stage, confirming the immuno-fluorescent studies, but could be displaced by an excess of the unlabelled ligand, providing evidence for the expression of insulin-binding receptors [6].

More recently, high resolution electron microscopic studies using colloidal gold-labelled insulin [11] have confirmed the earlier studies at the light microscopic level, and provided clear evidence that insulin binding to the preimplantation mouse embryo is receptor-mediated [20]. Figure 2 shows the distribution of gold-labelled insulin particles on the plasma membrane of mouse embryos at the 2-, 8-cell, morula and blastocyst stages.
Morphometric analysis of sections across whole embryos showed that the number of insulin receptors increased gradually during development, and markedly between the morula and blastocyst stages. As in the autoradiographic study, an approximately 75% displacement of labelled insulin by co-incubation in

excess unlabelled ligand confirmed the specificity of binding, and that it was receptor-mediated. Control embryos, incubated in gold-labelled BSA showed non-specific accumulation of gold particles in the zona pellucida, but there was no evidence of surface binding, or of internalization. Since the gold-insulin was stabilized by BSA, this result provided further evidence that uptake was specific to the hormone.

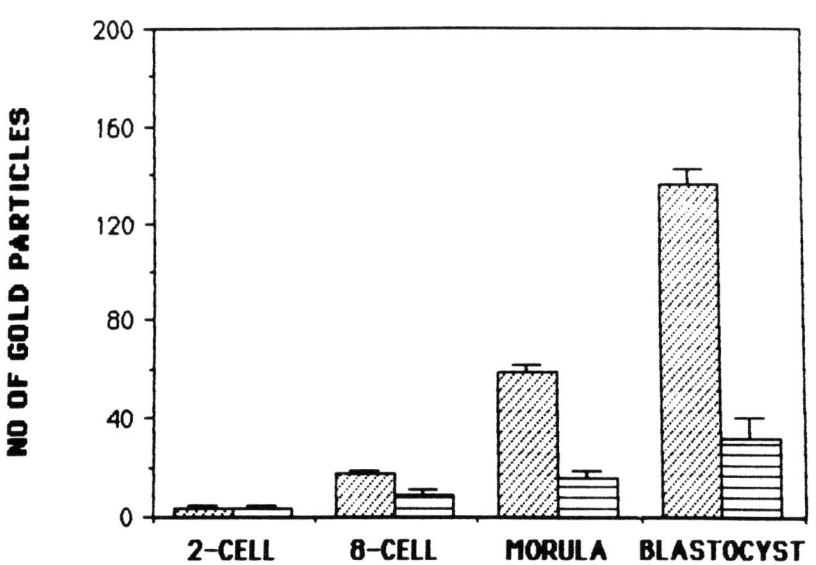

FIGURE 2. Specific binding of gold-labelled insulin (diagonal hatching) and its displacement by unlabelled ligand (horizontal hatching). The bar graphs show the number of gold particles bound to the plasma membrane of a developmental series of mouse preimplantation embryos. Values represent the number of gold particles per thin-sectioned embryo ± standard error of the mean.

Analysis of the uptake and endocytotic pathway of gold-labelled insulin by high-resolution electron microscopy revealed the following details: the first event is accumulation of the ligand along areas of

thickened plasma membrane, this is followed by
clustering within coated pits, and the ligand is
then translocated by means of coated and uncoated
vesicles, and accumulates in multivesicular bodies,
dense bodies, and endosomes (see Fig. 3).

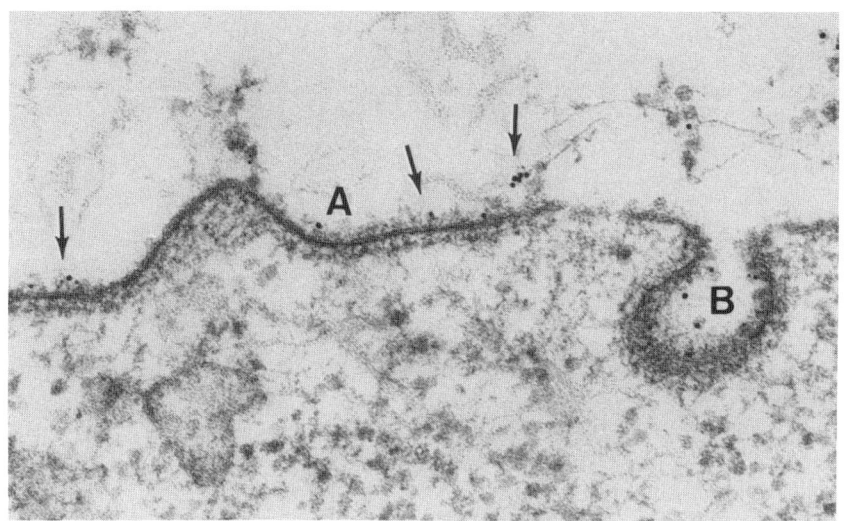

FIGURE 3. Electron micrographs illustrating binding
and internalization of gold-labelled insulin in the
blastocyst-stage mouse embryo. Gold-labelled
insulin (arrows) clustered over thickened areas of
trophectoderm plasma membrane (A) was internalized
in coated pits (B). Magnification = 85,000X

At the blastocyst stage, there was a
significant increase in receptor number. Although
the blastocyst is larger than the morula, and
trophectodermal cells are larger than those of the
morula, the increase in size and cell number was not
sufficient to account to the increase in receptor
number. In addition, morphometric analysis revealed
that approximately 21% of the occupied receptors on
the surface of the trophectodermal cells were found
in coated pits, indicating that uptake of insulin

was by a concentrative mode. In other cell types that have been studied using this technique, for example, the rat adipocyte, insulin uptake has been by means of non-coated pinocytotic invaginations [22].

A significant proportion of gold-labelled insulin in the blastocyst was internalized by cells of the polar trophectoderm, and was translocated through these cells in coated vesicles, and released into the space between the trophectoderm and cells of the ICM. The ligand bound subsequently to thickened areas of ICM cell membrane, and was internalized by coated pits. Gold particle counts on cells of the ICM in intact blastocysts were significantly lower than on trophectoderm cells. When isolated ICMs were examined for their ability to bind insulin, gold particle counts were similar. This difference may reflect the much lower concentration of insulin within the intact blastocyst.

A question of some importance is whether insulin remains immunologically intact after endocytosis in the embryo. To examine this, immunocytochemical studies were carried out to detect insulin as well as insulin receptors, in blastocysts. Studies of receptor expression using an anti-receptor antibody confirmed the labelled insulin studies. Further, receptors recognized by the antibody were localized in the same organelles as the gold-labelled insulin particles, with the exception of the dense bodies, suggesting that the receptors are absent, or degraded in these organelles. Immunologically-intact insulin could be detected both in the mural as well as the polar trophectoderm, and in cells of the ICM in blastocysts flushed from the reproductive tract and fixed immediately. Greater amounts of insulin could be detected in embryos that had been incubated in pharmacological amounts of insulin (200 ng/ml) for 2 hours, as shown in Fig. 4.

No insulin could be detected in blastocysts that were cultured overnight in medium lacking insulin, suggesting that preimplantation embryos do not synthesize insulin, and pointing to a maternal source. This was confirmed by immunocytochemical analysis of the reproductive tract, which showed insulin within the lumen. Our observations are in agreement with studies at the mRNA level in which the investigators were unable to detect insulin transcripts during preimplantation development [7].

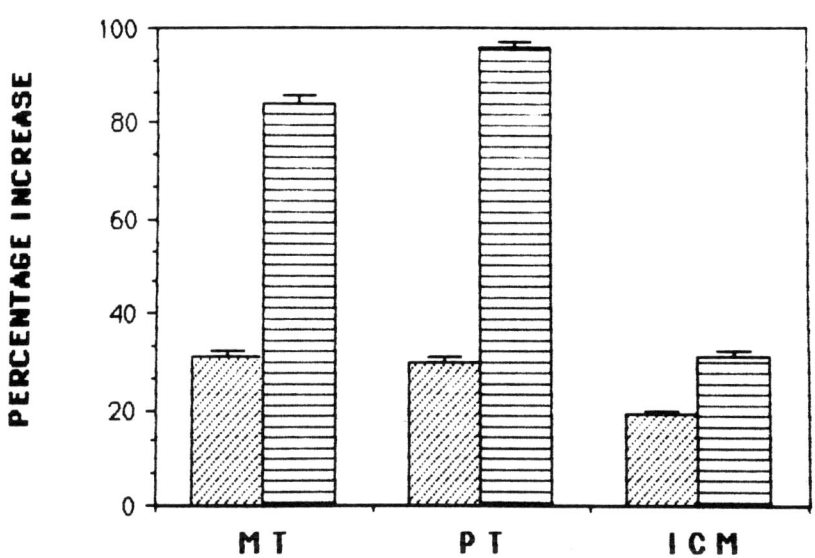

FIGURE 4. Detection of immunologically intact insulin in mouse blastocysts. Gold-labelled anti-insulin IgG was used to localize insulin. Values represent percent increase in gold particles in embryos cultured in 200 ng/ml insulin (horizontal hatching) over particle distribution in embryos flushed from the reproductive tract (diagonal hatching). Mural trophectoderm, MT; polar trophectoderm, PT; inner cell mass, ICM.

The results of the studies described above indicate that the early mouse embryo accumulates insulin from the maternal reproductive tract by means of receptor-mediated endocytosis. Further, the mode of uptake of the hormone is concentrative, and mediated by coated pits. These results suggest that insulin may play an important role in the early stages of development in the mouse.

The functional studies demonstrate clearly that insulin has a stimulatory effect on the rate of synthesis of DNA and RNA, showing insulin acting in a paracrine mode. However, in spite of the above observation, it is well known that mouse embryos can be induced to develop from the 2-cell stage to the blastocyst in the absence of insulin in the medium. Thus, insulin does not appear to be an absolute requirement for development. On the other hand, implantation is known to be the most vulnerable period of reproduction, and a good case can be made for insulin (and possibly other growth factors) acting in the preimplantation period to ensure rapid cellular growth and proliferation of the early postimplantation embryo.

ACKNOWLEDGMENTS

We are indebted to Neelima Shah and Maria Wikarczuk for skillful technical assistance.

REFERENCES

1. Kaye PL (1986). Metabolic aspects of the physiology of the preimplantation mouse embryo. In Rossant J, Pederson RA (eds): "Experimental Approaches to Mammalian Embryonic Development". Cambridge: Cambridge University Press. p. 267.
2. Mattson BM, Chambers SA, de Pablo F (1989). Comparative aspects of the conservation of insulin and the insulin receptor throughout evolution. In Rosenblum IY, Heyner S (eds): "Growth Factors in Mammalian Development". Boca Raton: CRC Press. In press.
3. Kaye PL (1988). Insulin stimulates protein synthesis in compacted mouse embryos. Endocrinology 116:261.
4. Gardner HG, Kaye PL (1984). Effects of insulin on preimplantation mouse embryos. Proc Aust Soc Reprod Biol 16:107.
5. Gardner D, Leese HJ (1988). The role of glucose and pyruvate transport in

regulating nutrient utilization by preimplantation mouse embryos. Development 104:423.
6. Mattson BM, Rosenblum IY, Smith RM, Heyner S (1988). Autoradiographic evidence for insulin and insulin-like growth factor binding to early mouse embryos. Diabetes 37:585.
7. Schultz GA, Rappolee DA, Pederson RA, Werb Z (1989). Changes in RNA and protein synthesis during development of the preimplantation mouse embryo. J Cell Biochem Suppl 13B:189.
8. Hogan BLM, Lacy E, Constantini F (1986). "Manipulating the Mouse Embryo". New York: Cold Spring Harbor Laboratory.
9. Whittingham DG (1971). Culture of mouse ova. J Reprod Fert 14:7.
10. Pursel VG, Wall RJ, Rexroad CE Jr, Hammer RE, Brinster RL (1985). A rapid whole mount staining procedure for nuclei of mammalian embryos. Theriogenology 24:687.
11. Smith RM, Goldberg RI, Jarett L (1988). Preparation and characterization of a colloidal gold-insulin complex with binding and biological activities identical to native insulin. J Histochem Cytochem 36:359.
12. Horisberger M, Vauthey M (1984). Labellling of colloidal gold with protein A, a quantitative study using D-lactoglobulin. Histochem 80:13.
13. Rosenblum IY, Mattson BM, Heyner S (1986). Stage-specific insulin binding in mouse preimplantation embryos. Dev Biol 116:261.
14. Nagarajan L, Anderson WB (1982). Insulin regulates the growth of F9 embryonal carcinoma cells apparently by acting through its own receptor. Biochem Biophys Res Comm 106:974.
15. Petruzelli LM, Ganguli S, Smith CJ, Cobb MH, Rubin CS, Rosen OM (1982). Insulin activates a tyrosine specific protein kinase in extracts of 3T3-LI adipocytes and human placenta. Proc Natl Acad Sci USA 79:6792.
16. Kahn CR, White MF (1988). The insulin receptor and the molecular mechanism of

insulin action. J Clin Invest 82:1151.
17. Ullrich A, Bell JR, Chen EY, Herrera R, Petruzzelli LM, Dull TJ, Gray A, Coussens L, Liao Y, Tsubokawa M, Mason A, Seeburg P, Grunfeld C, Rosen O, Ramachandran J (1985). Human insulin receptor and its relationship to the tyrosine kinase family of oncogenes. Nature 313:756.
18. Ebina Y, Ellis L, Jarnagin K, Edery M, Graf L, Clauser E, Ou J, Masiarz F, Kan YW, Goldfine I, Rother RA, Rutter WJ (1985). The human insulin receptor cDNA:the structural basis for hormone-activated transmembrane signalling. Cell 40:747.
19. Czech MP (1984). New perspectives on the mechanism of insulin action. Rec Prog Horm Res 40:347.
20. Fleming TP, Pickering SJ (1985). Maturation and polarization of the endocytotic system in outside blastomeres during mouse preimplantation development. J Embryol exp Morph 89:175.
21. Heyner S, Rao LV, Jarett L, Smith RM (1989). Preimplantation mouse embryos internalize maternal insulin via receptor-mediated endocytosis:pattern of uptake and functional correlations. Dev Biol In press.
22. Smith RM, Jarett L (1988) Receptor mediated endocytosis and intracellular processing of insulin:ultrastructural and biochemical evidence for cell-specific heterogeneity and distinction from non-hormone ligands. Lab Invest 58:613.

EPIDERMAL GROWTH FACTOR AND PREGNANCY IN THE MOUSE[1]

Y.M. Huet-Hudson[2], G.K. Andrews and S.K. Dey

Departments of Obstetrics-Gynecology and Physiology (YMH & SKD)
and Department of Biochemistry and Molecular Biology (GKA)
Ralph L. Smith Research Center
University of Kansas Medical Center
Kansas City, Kansas 66103

ABSTRACT

Uterine cell type-specific expression of EGF was analyzed during the peri-implantation period by localization of EGF mRNA and EGF protein using in situ hybridization and immunocytochemistry, respectively. EGF gene expression was restricted to the luminal and glandular epithelial cells where it occurred transiently, with EGF mRNA and protein being detected late on proestrus and early on day 1 of pregnancy. Neither were detected late on day 1 or on days 2 and 3. Although EGF immunostaining was again detected on day 4, and the staining appeared to be at or near the apical border of the luminal epithelium, EGF mRNA was not detected. On day 5, neither EGF mRNA nor protein was detected. EGF gene expression in epithelial cells of the mouse uterus was enhanced by estrogen, thus preovulatory estrogen may regulate uterine EGF levels during late proestrus. Long-term sialoadenectomy did not alter the timing or specificity of EGF gene expression in the uterus. Likewise, EGF levels in the plasma were not reduced, and implantation and pregnancy maintenance were not adversely affected. The results suggest that uterine expression of the EGF gene is tightly regulated in a cell type-specific and temporal manner during pregnancy. Thus, uterine EGF may have a role during pregnancy, but a role for submandibular gland EGF can be excluded.

---

[1]This work was supported by NIH grants (HD 12304 and ES 04725). YMH was supported by NSF and Ford Foundation Fellowships.
[2]Present address: Monsanto Company, AA4C, 700 Chesterfield Village Parkway, St. Louis, Missouri 63198.

INTRODUCTION

Epidermal growth factor (EGF) is a small polypeptide (53 amino acids) with well documented effects on cell division and differentiation. The mature EGF molecule is formed by proteolytic processing of a large precursor protein (~130 kd) termed preproEGF. The organs having the highest levels of EGF are mouse submandibular gland and kidney, but low levels have been detected in a variety of other organs (1,2). Processed EGF is most abundant in the submandibular gland, whereas in the kidney preproEGF is most abundant (2). EGF exerts its effects via specific receptor binding. The gene encoding the EGF receptor is homologous with the v-erb B oncogene (3-5). The EGF receptor binds to either EGF or TGF-α and induces EGF-like effects (6,7). Although the definitive physiologic functions of EGF have not been determined, this growth factor has been shown to influence cell proliferation (8-10), steroid synthesis (11,12), uterine smooth muscle contraction (13), prostaglandin synthesis (14-16), PI turnover (17), intracellular rise in Ca^{++} (17,18) and induction of c-myc (19).

Recently it was reported that the concentration of EGF in the submandibular gland and plasma of C3H/HeN mouse increased during pregnancy. Apparently, pregestational sialoadenectomy prevented the rise in plasma EGF level, and increased the incidence of abortion; an effect corrected by administration of exogenous EGF (20). These investigators also noted that the administration of antisera to EGF in sialoadenectomized mice increased the percentage of abortions (20). These results suggested that EGF, of submandibular gland origin, functions in an endocrine manner and has a significant role in pregnancy outcome. On the other hand, immunoreactive EGF (21) and EGF receptors (22,23) increase in the rat uterus after estrogen treatment and during the peri-implantation period (24). Therefore, it has been suggested that EGF may be produced in the uterus and modulates uterine and/or embryonic function via an autocrine/paracrine mechanism. In the present investigation, cell type-specific localization of EGF and EGF mRNA was determined by immunocytochemistry and in situ hybridization during the peri-implantation period. Furthermore, to reevaluate the role of circulating EGF of submandibular gland origin, long-term effects of sialoadenectomy on uterine distribution and plasma level of EGF, as well as on implantation and the success of pregnancy were investigated.

MATERIALS AND METHODS

Animals

Female CD-1 mice (48 days old, Charles River Laboratories, NJ) were used. Females were mated with fertile males of the same strain. The morning of finding a vaginal plug was defined as day 1 of pregnancy. For immunocytochemical localization of EGF and in situ hybridization of EGF mRNA, mice were killed every two hours beginning at 1800 h on the day of proestrus until 1000 h on day 1 of pregnancy and at 0830-0900 h on days 2-5. Pseudopregnant mice, produced by mating with vasectomized males, were killed on day 5.

To determine the effects of sialoadenectomy on pregnancy and plasma levels of EGF, submandibular glands from virgin females were surgically removed bilaterally under avertin anesthesia. Mice were rested three weeks before being used in the following experiments. Groups of sham-operated and sialoadenectomized mice were mated with males. Some females were killed on day 1 for immunocytochemical localization of EGF or on day 5 to check for implantation sites, as determined by intravenous injections of Chicago Blue B (25). The other females were maintained until term and litter size was recorded. At the end of experiments, blood was either collected from the trunk after exsanguination of the head (under avertin anesthesia) or from the abdominal aorta using a syringe. Plasma was assayed for EGF by radioimmunoassay.

Immunolocalization of EGF

Uteri were excised, cleaned of fat, cut into 4 mm pieces and rapidly frozen in liquid Freon. Frozen uteri were sectioned at 8 μM and mounted on poly-L-lysine coated slides. Sections were vacuum dried, and then fixed in cold acetone for 10 min prior to use. Sections were incubated in phosphate buffered saline (PBS) for 20 min, and then they were incubated in primary antibody (rabbit polyclonal antisera against mouse or rat EGF, kindly provided by Dr. R. DiAugustine, NIEHS, NC, and Dr. K. Mayo, Temple University, PA) at 1:500 dilution for 2 h at 25°C. Immunocytochemical staining was performed using a Zymed Histostain-SP kit for rabbit primary antibody (Zymed laboratories, San Francisco, CA) which utilizes a biotinylated secondary antibody, a horseradish peroxidase-strepavidin conjugate and a substrate chromogen mixture (26). Endogenous peroxidase activity was blocked by exposing sections to 0.23% periodic acid (Sigma Chemical Co., St. Louis, MO) in PBS for 35 seconds following secondary antibody incubation (27). Sections were counterstained with hematoxylin, mounted and examined under brightfield. Control experiments included incubation of sections with preimmune serum, in the absence of primary antibody or with primary antibody neutralized with excess antigen.

In situ hybridization of EGF mRNA

The procedures for in situ hybridization have been adapted from methods published by Angerer and associates (28-31). The details of this technique have been described previously (32). Hybridization was performed on paraformaldehyde perfused fixed uterine sections (7 μM) using a ^{35}S labelled RNA probe complementary to the 3'region of EGF mRNA (provided by Dr. G.I. Bell, Northwestern University, Chicago, IL) or a ^{32}P-labelled oligodeoxyribonucleotide probe complementary to 40 nucleotides of the EGF coding region of EGF mRNA. Sections of male mouse submandibular glands were mounted onto the same microscope slide as positive controls.

Radioimmunoassay of EGF

Plasma was assayed directly without extraction using a kit (Diagnostic Systems Laboratories,Inc., Webster, TX) which utilized a double antibody technique (33). (^{125}I)Iodo-EGF was used as tracer. Extracts of male mouse submandibular glands were used as positive controls. The lowest detectable level of EGF that could be distinguished from the 0-standard was 10 pg/tube at the 95% confidence limit. The intra- and interassay coefficients of variation did not exceed 8% and 6% respectively.

RESULTS

Immunolocalization of EGF

No non-specific staining was observed when sections were incubated with preimmune serum, in the absence of primary antibody or with primary antibody neutralized with excess antigen (data not shown). As a positive control, cells in the submandibular gland, but not parotid gland, stained positively for EGF (Figure 1a). The localization of this protein in the uterus was limited exclusively to the luminal and glandular epithelia under all conditions examined. At 1800 h on proestrus, no immunoreactive EGF was detected, but between 2200 and 2400 h the cytoplasm of luminal and glandular cells stained positively (data not shown). On day 1 of pregnancy, the staining was visible at 0 h, and continued to be present at 0200, 0400, 0600 and 0800 h (localization for 0600 is shown, Figure 1d). Day 1 (0800 h) uterine sections from long-term sialoadenectomized mice also showed similar EGF localization (data not shown). EGF immunostaining was not detected by 1000 h on day 1, or on days 2 and 3 (Figure 1e). Interestingly, this immunostaining was again detected on day 4, but exclusively at or near the apical border of the luminal epithelium (Figure 1f). Although this staining had almost disappeared on day 5 of

EGF and Pregnancy 129

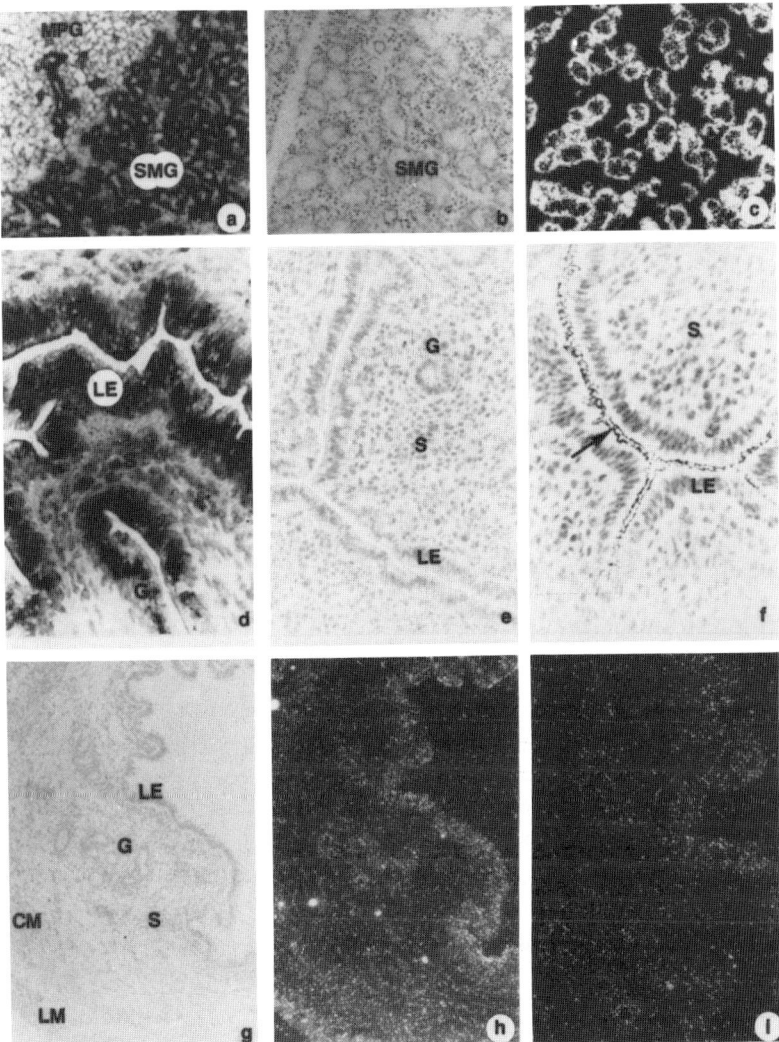

Figure 1. Immunocytochemical localization of EGF and in situ hybridization of EGF mRNA in male mouse salivary glands and pregnant uterus. **Immunolocalization**: (a) male submandibular gland (SMG) and parotid gland (MPG), X100; (d) day-1 uterus (0600 h); (e) day-3 uterus, and (f) day-4 uterus (staining indicated by an arrow), X200 (bright field). Note that the parotid gland and day-3 uterus did not show immunostaining. **In situ hybridization**: (b) SMG (bright field); (c) SMG (dark field, ^{35}S-cRNA probe and 4 day autoradiographic exposure, X100); (g) day-1 uterus (bright field); (h) day-1 uterus (dark field, X100) and (i) day-1 uterus (dark field, ^{35}S-cRNA probe and 14 day autoradiographic exposure, X200). Under the dark field, areas of dense silver grains indicate sites of probe hybridization. LE = luminal epithelium; G = gland; S = stroma; CM = circular muscle; and LM = longitudinal muscle.

pregnancy, it was retained on day 5 of pseudopregnancy (data not shown).

In situ hybridization of EGF mRNA

EGF mRNA was detected in a cell type-specific manner which colocalized with EGF protein both in the submandibular gland and day 1 uterus (Figure 1b,c,g,h,i). Two specific hybridization probes (^{35}S-labelled cRNA and ^{32}P-labelled oligodeoxyribonucleotide) showed similar localization. Although EGF was not detected at 2000 h on proestrus, EGF mRNA was detectable at this hour (data not shown). Thereafter, EGF mRNA was detected at each time point examined on day 1, except at 1000 h. EGF mRNA was not detected on the morning of days 2 to 5 of pregnancy, and no other times of day were examined.

Effect of long-term sialoadenectomy on pregnancy

Long-term sialoadenectomy did not cause adverse effects on either implantation or the entire course of pregnancy in CD-1 mice. All mice in sham-operated and sialoadenectomized groups had implantation sites when examined on day 5, and the mean number of implantation sites was statistically the same for both of these groups (Table 1). The number of sialoadenectomized mice completing term pregnancy was not different from the sham-operated controls, and the litter size was similar in both of these groups (Table 2).

Effect of long-term sialoadenectomy on plasma EGF

The plasma levels of EGF varied depending upon the site and the procedure of blood collection. When blood was collected from the trunk region, EGF levels were more than 800-fold higher in the sham-operated mice as compared to those of sialoadenectomized animals. On the other hand, levels were quite low and not different between the two groups, when blood was collected from the abdominal aorta (Table 3).

DISCUSSION

The results of the present investigation clearly suggest that sialoadenectomy does not have apparent deleterious effects on the initiation and maintenance of pregnancy in CD-1 mice. The suggestion by Oka and associates that the submandibular gland is the major source of circulating EGF (20,34,35) is in contrast to our present findings, and earlier findings of others (36,37). The reported decline in plasma EGF levels following removal of the submandibular glands (20,34,35) raises the question as to how EGF from the submandibular glands enters into the circulation, since EGF is apparently an exocrine product (38). Oka

Table 1 Effect of long-term sialoadenectomy on implantation in the mouse.

Conditions	No. of Mice	No. of Mice with I.S.	Mean No. of I.S.
Sham Control	6	6	10.2 ± 1.2
Sialoadenectomy	15	15	7.9 ± 0.8

Results shown are means (± SEM). Implantation sites were delineated by an intravenous injection of 0.1 ml of 1% Chicago Blue B (Sigma Chemical Co., St. Louis, MO) 15 min before sacrifice under anesthesia. Discrete blue bands along the uterine horns indicated implantation sites (I.S.).

* * * * * * *

Table 2 Effect of long-term sialoadenectomy on pregnancy outcome in the mouse.

Conditions	No. of Mice	No. of Mice completed pregnancy	Mean litter size
Sham Control	8	8	8.9 ± 0.4
Sialoadenectomy	12	11	7.8 ± 0.4

Results shown are means (± SEM). Litter size was recorded and pups were checked for any visible abnormalities 24 h after they were born. No apparent abnormalities were observed.

* * * * * * *

Table 3 Effect of long-term sialoadenectomy on plasma EGF levels in the mouse.

Conditions	Trunk plasma (ng/ml)	Abdominal aorta plasma (pg/ml)
Sham Control	304.1 ± 113.8 (7)	110.41 ± 33.1 (5)
Sialoadenectomy	0.37 ± 0.04 (9)	112.56 ± 30.4 (5)

Values are means (± SEM). Blood was collected from lactating mice 5 days after they delivered.

and his group suggested that EGF could be reabsorbed from the digestive tract into the circulatory system (35). However, it is clear from these studies that the method of collection of blood can drastically influence the levels of EGF detected in the sample. Blood from the trunk region, especially in animals with intact salivary glands, erroneously gives rise to high levels of plasma EGF. This is probably due to anatomical location of the submandibular glands. Any manipulation in collecting blood from that region without extreme precaution may contaminate the blood samples with salivary gland EGF. This appears to be the case since blood samples obtained from the abdominal aorta of both the sialoadenectomized and sham-operated animals showed the same low level of plasma EGF. These findings bring into question models which predict an endocrine function for submandibular EGF.

The question of EGF having a role in pregnancy is an entirely different issue. EGF has been proposed to influence various cellular functions in an autocrine/paracrine manner. Therefore, it is possible that estrogen-progesterone modulation of proliferation and differentiation of various cell-types of the uterus and embryo is mediated by local release of EGF singularly or in combination with other growth factors. Indeed, EGF receptors are present in the uterus and the embryo (22-24,39), and this growth factor stimulates uterine epithelial cell proliferation in vitro (40). The finding reported here that both EGF mRNA and protein are present in the uterine epithelia, strongly supports the concept that EGF is locally produced in the uterus. Although the uterus is apparently another organ that produces this growth factor, EGF is in significantly low abundance in this organ as compared to submandibular glands. However, removal of the submandibular glands has no apparent effects on the initiation and maintenance of pregnancy.

The expression of the EGF gene in the uterus appears to be transient and tightly regulated by the endocrine state of the female. Preovulatory estrogen likely induces this expression during proestrus and early on day 1 of pregnancy. A recent report also provides evidence that estrogen stimulation results in expression of EGF mRNA in the immature mouse uterus, and suggests that this growth factor in the uterus is primarily in the prepro form (33). On the other hand, our recent work indicates that estrogen-dependent EGF in the adult mouse uterus is mainly in the processed form, as it is in the submandibular gland (to be published).

The functional significance of EGF in the uterus is not known. Because appearance of uterine EGF is dependent on estrogen stimulation, it is suggested that EGF participates in estrogen-induced uterine growth. This could then suggest that uterine growth during late proestrus or day 1 of pregnancy involves EGF as one of the mediators of estrogen action. The exact nature, function and source of the immunoactive EGF that was localized at or near the apical border of the luminal epithelium on day 4 of pregnancy are not clearly understood. It seems likely that this EGF results from transient uterine expression of this gene during the night of day 3 or early on day 4, but that remains to be determined. Also, whether this EGF is membrane bound (preproEGF), secreted into or imported into the uterine lumen is unknown. The disappearance of this immunoreactive EGF following implantation on day 5 of pregnancy, in contrast to its persistence on the same day in the pseudopregnant mouse (data not shown), suggests a possible functional significance in the events of implantation.

The results described here, and to be published elsewhere, provide evidence that specific cell-types in the adult mouse uterus can synthesize EGF or EGF-like proteins under the influence of ovarian steroids. However, the physicochemical properties and possible growth factor activity of this uterine immunoreactive EGF remain to be delineated as do the functional roles of these proteins.

ACKNOWLEDGEMENTS

We thank Katherine Schott for her technical assistance and Linda Hicks for her help in preparation of the manuscript.

REFERENCES

1. Cohen S (1962). Isolation of a mouse submaxillary gland protein accelerating incisor eruption and eyelid opening in the new-born animal. J Biol Chem 237:1555.
2. Rall LB, Scott J, Bell GI, Crawford RJ, Penschow JD, Niall HD, Coghlan JP (1985). Mouse prepro-epidermal growth factor synthesis by the kidney and other tissues. Nature 313:228.
3. Defize LHZ, De Laat SW (1986). Structural and functional aspects of epidermal growth factor (EGF) and its receptor. In Gispen WH, Routtenberg A (eds):" Progress in Brain Research, Vol.69,"Elsvier Publishers B.V.(Biomedical Division), New York, p 169.
4. Downward J, Yarden Y, Mayes E, Scrace G, Totty N, Stockwell P, Ullrich A, Schlessinger J, Waterfield MD (1984). Close similarity of epidermal growth factor receptor and v-erb-B oncogene protein sequences. Nature 307:521.
5. Ullrich A, Coussens L, Hayflick JS, Dull TJ, Gray A, Tam AW, Lee J, Yarden Y, Libermann TA, Schlessinger J, Downward J, Mayes ELV, Whittle N, Waterfield MD, Seeburg PH (1984). Human epidermal growth factor receptor DNA sequence and

aberrant expression of the amplified gene in A431 epidermal carcinoma cells. Nature 309:418.
6. DeLarco JE, Todaro GJ (1980). Transforming growth factors produced by retrovirus-transformed rodent fibroblasts and human melanoma cells: Amino acid sequence homology with epidermal growth factor. Proc Natl Acad Sci 80:4684.
7. Pike LJ, Marquardt H, Todaro GJ, Gallis B, Casnellie JE, Bornstein P, Krebs EG (1982). TGF and EGF stimulate the phosphorylation of synthetic tyrosine-containing peptide in a similar manner. J Biol Chem 257:14628.
8. Carpenter G, Cohen S (1978). Epidermal Growth Factors. In Litwack G (ed): "Biochemical Actions of Hormones Vol.5," New York: Academic Press, p 203.
9. Carpenter G, Cohen S (1979). Epidermal growth factor. Ann Rev Biochem 48:193.
10. Hollenberg MD (1979). Epidermal growth factor-urogastrone, a polypeptide acquiring normal status. Vitam Horm 37:69.
11. Jones PBC, Welsh TH, Jr, Hsueh AJW (1982). Regulation of ovarian progestin production by epidermal growth factor in cultured granulosa cells. J Biol Chem 257:11268.
12. Ascoli M, Euffa J, Segaloff DL (1987). Epidermal growth factor activates steroid biosynthesis in cultured leydig tumor cells without affecting the levels of cAMP and potentiates the activation of steroid biosynthesis by chorionic gonadotropin and cAMP. J Biol Chem 262:9196.
13. Gardner RM, Lingham RB, Stancel GM (1987). Contractions of the isolated uterus stimulated by epidermal growth factor. FASEB J 1:224.
14. Chiba T, Hirata T, Kadowaki S, Matsukura S, Fujita T (1982). Epidermal growth factor stimulates prostaglandin E release from isolated perfused rat stomach. Biochem Biophys Res Commun 105:370.
15. Mitchell MD (1987). Epidermal growth factor actions on arachidonic acid metabolism in human amnion cells. Biochim Biophys Acta 928:240.
16. Yokota K, Kusaka M, Ohshima T, Yamamoto S, Kurihara N, Yoshino T, Kumegawa M (1986). Stimulation of prostaglandin E2 synthesis in cloned osteoblastic cells of mouse (MC3T3-E1) by epidermal growth factor. J Biol Chem 261:15410.
17. Sawyer ST, Cohen S (1981). Enhancement of calcium uptake and phosphatidylinositol turnover by epidermal growth factor. Biochemistry 20:6280.
18. Moolenaar WH, Aerts RJ, Tertoolen LGJ, De Laat SW (1986). The epidermal growth factor-induced calcium signal in A431 cells. J Biol Chem 261:279.

19. Reuse S, Roger PP, Vassart G, Dumont JE (1986). Enhancement of c-myc mRNA concentration in dog thymocytes initiating DNA synthesis in response to thyrotropin, forskolin, epidermal growth factor and phorbol myristate ester. Biochem Biophys Res Commun 141:1066.
20. Tsutsumi O, Oka, T (1987). Epidermal growth factor deficiency during pregnancy causes abortion in mice. Am J Obstet Gynecol 156:241.
21. Gonzalez F, Laksmanan J, Hoath S, Fisher DA (1984). Effect of oestradiol-17β on uterine epidermal growth factor concentration in immature mice. Acta Endocrinol 105:425.
22. Mukku VR, Stancel GM (1985). Regulation of epidermal growth factor receptor by estrogen. J Biol Chem 260:9820.
23. Lingham R, Stancel GM, Loose-Mitchell DS (1988). Estrogen induction of the epidermal growth factor receptor mRNA. Mol Endocrinol 2:230.
24. Chakraborty C, Tawfik O, Dey SK (1988). Epidermal growth factor binding in rat uterus during the peri-implantation period. Biochem Biophys Res Commun 153:564.
25. Psychoyos A (1973). Endocrine control of egg implantation. In Greep RO, Astwood EG, Geiger SR (eds):"Handbook of Physiology," Washington, D.C.: American Physiological Society, p 187.
26. Hsu S-M, Raine L (1984). The use of avidin-biotin-peroxidase complex (ABC) in diagnostic and research pathology. In DeLellis RA (ed):"Advances in Immunochemistry," New York: Mason Publishing USA Inc, P 31.
27. Kelly J, Whelan CA, Weir DG, Reighery C (1987). Removal of endogenous peroxidase activity from cryostat sections for immunoperoxidase visualization of monoclonal antibodies. J Immun Methods 96:127.
28. Angerer LM, Angerer RC (1981). Detection of poly-A+ RNA in sea urchin eggs and embryos by quantitative in situ hybridization. Nucl Acids Res 9:2819.
29. Angerer LM, Deleon DV, Angerer RC, Showman RM, Wells DE, Raff RA (1984). Delayed accumulation of maternal histone mRNA during sea urchin oogenesis. Dev Biol 102:477.
30. Cox KH, Deleon DV, Angerer LM, Angerer RC (1984). Detection of mRNAs in sea urchin embryos by in situ hybridization using asymmetric RNA probes. Dev Biol 101:485.
31. Deleon DV, Cox KH, Angerer LM, Angerer RC (1983). Most early-variant histone mRNA is contained in the pronucleous of sea urchin eggs. Dev Biol 100:197.
32. De SK, Dey SK, Andrews GK (1989). Cell-specific metallothionein gene expression in mouse decidua and placenta. Biol Reprod (submitted).

33. DiAugustine RP, Petrusz P, Bell GI, Brown CF, Korach KS, McLachlan JA, Teng CT (1988). Influence of estrogens on mouse uterine epidermal growth factor precursor protein and messenger ribonucleic acid. Endocrinology 122:235.
34. Kurachi H, Oka T (1985). Changes in epidermal growth factor concentrations of submandibular gland, plasma and urine of normal and sialoadenectomized female mice during various reproductive stages. J Endocrinol 106:197.
35. Tsutsumi O, Kurachi H, Oka T (1986). A physiological role of epidermal growth factor in male reproductive function. Science 233:975.
36. Byyny RL, Orth DN, Cohen S, Doyne ES. (1974). Epidermal growth factor: Effects of androgens and adrenergic agents. Endocrinology 95:776.
37. Barka, T (1980). Biologically active polypeptides in submandibular glands. J Histochem Cytochem 28:836.
38. Murphy RA, Watson AY, Metz J, Forssmann WG (1980). The mouse submandibular gland: An exocrine organ for growth factors. J Histochem Cytochem 28:890.
39. Adamson ED, Meek J (1984). The ontogeny of epidermal growth factor receptors during mouse development. Dev Biol 103:62.
40. Tomooka Y, DiAugustine RP, McLachlan JA (1986). Proliferation of mouse uterine epithelial cells in vitro. Endocrinology 118:1011.

INSULIN-LIKE GROWTH FACTOR BINDING PROTEIN AND
PREGNANCY: REGULATION AND FUNCTION IN THE PRIMATE[1]

A.T. Fazleabas[2], H.G. Verhage[2] and S.C. Bell[3]

[2]Department of Obstetrics and Gynecology
University of Illinois Chicago, IL 60612 and
[3]Departments of Obstetrics and Gynecology and
Biochemistry University of Leicester
Leicester LE2 7LX, U.K.

Background

Morphological studies in the primate have demonstrated association between the developing trophoblast and decidualized endometrium. However, the potential regulation of these interactions by conceptus and/or maternal secretory products have yet to be elucidated. Studies of this nature in the primate have been hampered up to the present due to the fact that a) they are not morally or ethically permissible in the human, b) lack of an appropriate non-human primate model and c) the unavailability of appropriate reagents to specific endometrial secretory proteins. The latter two problems have been resolved to some extent. First, our studies have demonstrated that baboon uterine endometrium secretes a number of proteins that are biochemically and immunologically similar to those secreted by the human endometrium (1,2). Second, monoclonal and polyclonal antibodies and complimentary deoxyribonucleic acid (cDNA) sequences have become available to the two major secretory proteins of the human endometrium and decidua (3-8).

---

[1]Work in the authors laboratories have been supported by NIH grants HD 21991 and HD 20571 (ATF, HGV) and the MRC and Wellcome Trust (SCB).

Pregnancy associated endometrial α_2-globulin (α_2-PEG), a glycosylated β-lactoglobulin homologue, is quantitatively the major secretory protein of the human glandular epithelium during the luteal phase of the menstrual cycle (9). Pregnancy associated endometrial α_1-globulin (α_1-PEG), an insulin-like growth factor binding protein (IGF-BP), is a minor secretory product during the luteal phase of the human menstrual cycle (10) but becomes the major secretory protein during the first trimester of pregnancy (9). These two proteins are analogous to placental proteins 14 and 12 (11). Comparative studies on the baboon endometrium during the menstrual cycle and pregnancy indicated that an immunologically similar protein to α_2-PEG is not synthesized by the uterine endometrium of this non-human primate (1). However, IGF-BP shows a similar synthetic profile in the baboon, with the exception that during the menstrual cycle this protein is predominantly localized in the glandular epithelium of the baboon in contrast to its primary localization in the predecidualized stromal cells of the human endometrium (12,13).

There are at least two distinct classes of IGF-BP. In plasma, IGF's are found bound to a complex of M_r 150,000 which consists of a growth hormone (GH)-dependent protein of M_r 53,000 and an acid-labile protein component (14,15). The GH-independent class of IGF-BP has a M_r of between 30-40,000 and is synthesized by a number of cell lines (16,17) and tissues, particularly the secretory phase endometrium and decidual tissue (13, 18-20). This protein is also present in high levels in physiological fluids such as amniotic fluid and milk (10,21). This review will focus on the properties of the GH-independent IGF-BP synthesized and secreted by the primate uterine endometrium and decidua and discuss its potential role in pregnancy.

Synthesis and Secretion IGF-BP

The close association between the developing trophoblast and decidualized endometrium together with evidence for insulin-like growth factor (IGF) production by placental tissues (22), suggest that decidual IGF-BP in the baboon and human may regulate feto-placental growth. The IGF's are a family of polypeptides, related to insulin, which stimulate growth and differentiation in a number of tissues during fetal development and post-natally (23, 24). These peptides (IGF I & II) bind to specific receptors on the cell surface and are usually found complexed to soluble binding proteins (15). These IGF-BP's bind IGF's with comparable affinity to that of their receptors and are able to modulate IGF action (25,26).

Stromal cells undergo a process of decidualization occluding intercellular spaces, particularly around blood vessels (27). It is thought that decidualization may be a progesterone dependent process (28). In the human, predecidualized stromal cells are found around spiral arteries in the luteal phase of the menstrual cycle and fully decidualized cells are formed around the twelfth day of pregnancy (29). Decidualization begins around day 16 to 17 of pregnancy in the baboon (30). Of particular interest, is the role the decidual cell may play in implantation and early pregnancy since it synthesizes and secretes the GH-independent IGF-BP as its major secretory product. Although the synthesis and secretion of this protein is a property of many cell types, none appear to secrete this protein as abundantly as the primate decidual cells (9,18). Comparative studies on human and baboon decidual IGF-BP indicate that the protein from both species is biochemically and immunologically similar. The secreted molecule is an acidic protein with a M_r 32-33,000, which migrates as a fused doublet on two-dimensional gels and binds IGF-I with a Kd of 1.14-1.83 nM (18).

IGF-BP constitutes a minor in vivo secretory product during the menstrual cycle of the human, but its secretion increases during the first trimester of pregnancy to reach peak values in amniotic fluid around weeks 20 to 24 (9,28). During the menstrual cycle in the human, IGF-BP is immunocytochemically localized to predecidualized stromal cells associated with spiral arteries of the endometrium, with occasional staining in the glandular epithelia (12). During pregnancy, IGF-BP is immunocytochemically localized to the decidua compacta while staining is absent in the placenta (28). In the pregnant baboon, IGF-BP is also localized to the decidual tissue (18). However, in the non-pregnant baboon, immunocytochemical staining with the same antibody localized IGF-BP to the endometrial epithelium of deep glands during the luteal stage of the cycle (13). It is not certain what factors regulate IGF-BP synthesis and secretion in the baboon endometrium, nor what may be regulating the switch in the cell-type which secretes IGF-BP during pregnancy. In vitro secretion studies using endometrial explant cultures obtained from ovariectomized, steroid-treated baboons showed that while neither progesterone nor estrogen alone causes secretion of IGF-BP, progesterone treatment after estrogen priming does result in IGF-BP secretion (13). Immunocytochemical evaluation of these endometria show that it is the glandular epithelium which synthesizes IGF-BP in these steroid-treated animals since staining was present in the glands and absent in the stroma (13). While this does indicate some regulation of secretion by progesterone in the endometrium of non-pregnant intact or steroid-treated animals, the hormone(s) or factor(s) which causes the change from epithelial to stromal (decidual) production of IGF-BP in the pregnant baboon remains unknown.

In addition to the difference in immunolocalization of the protein during the menstrual cycle between the two species, monoclonal antibodies with different epitope specificities also stain differently. Three monoclonal antibodies have been generated towards human IGF-BP, namely B2H10, C3H12 and C4H11 (19). All three monoclonal antibodies cross react with pregnant and non-pregnant human tissue and pregnant baboon tissues, however C4H11 does not cross react with the baboon protein during the menstrual cycle (13). While the functional significance of this observation is unknown, it is apparent from the antibody characterization studies that C4H11 binding to IGF-BP is inhibited when the BP is saturated with IGF-I (19). It has been suggested that IGF-BP in the human may play a role in either regulating trophoblast growth or permitting proliferating stromal fibroblasts to respond to endogenous IGF (31,32). Recent evidence suggests that the IGF-I/IGF-BP complex is a better mitogen than free IGF-I (33). Since implantation appears to be more stringently controlled in the baboon following the initial wave of rapid trophoblast invasion, one might suggest that IGF-BP may play a role in regulating trophoblast penetration. Perhaps, the IGF-BP complexed with IGF-I and localized in the deep glandular epithelium acts as a mitogen on conceptus tissues and facilitates trophoblast penetration and contact with the maternal vasculature. Following implantation, the switch in the site of synthesis in the baboon and the continued rise in IGF-BP synthesis by the decidual cells in the human may then be responsible for stromal proliferation and control of trophoblast invasiveness. The reported stimulatory and inhibitory effects of IGF-BP on the mitogenic action of IGF-I are discussed later in this review.

Cloning and Sequencing of cDNA to IGF-BP

The cDNA clones for the GH-dependent (BP-53; 34) and the GH-independent (BP-25) IGF-BP's (5-7) have been isolated and the amino acid sequence determined from the neuclotide sequence. BP-53 consists of a translated core protein of 264 amino acids and has three potential glycosylation sites (34). BP-25 consists of 234 amino acids and is not glycosylated (7). Comparison of the cDNA sequences of BP-53 and BP-25 reveal a 33% overall homology, with 100% homology in the cysteine-rich domains (34), thus suggesting that these two BP's are the products of different chromosomal genes with a common ancestral gene (7,34). The cDNA for the baboon GH-independent IGF-BP has also been recently isolated from a lambda zap II baboon decidual library. Partial nucleotide sequencing indicates that in addition to baboon IGF-BP being biochemically and immunologically similar to human, the DNA sequences for the two IGF-BP's are also virtually identical (A.T. Fazleabas, K. Donnelly and R.C. Jaffe, unpublished observations).

Although the IGF-binding site on the BP has not been clearly delineated, the M_r 21,000 fragment of the BP isolated from amniotic fluid and containing the N-terminal region has been shown to bind IGF (35). The 100% homology in the cysteine-rich domains near the N and C-terminus between BP-53 and BP-25 may be functionally important for ligand binding, similar to a variety of hormone receptors (36-38). Another important structural feature revealed by DNA sequencing of BP-25 is the presence of an Arg-Gly-Asp (RGD) tripeptide near the C-terminus (5,7). This consensus sequence appears to be essential for extracellular proteins which bind to the integrin class of receptors on cell membranes (39). Preliminary experiments have demonstrated that a synthetic RGD tripeptide is capable of inhibiting the binding of amniotic fluid IGF-BP to cell membranes (5). The relevance of this observation may relate to the demonstrated ability of IGF-BP to both inhibit and enhance the action of IGF-1 on target cells (40; see following section). It is therefore conceivable that the N-terminal fragments which retain their IGF binding ability (35) but do not process the C-terminus RGD tripeptide would inhibit the action of IGF-1, while the presence of the RGD tripeptide on the molecule would enhance its membrane-mediated stimulatory action.

Northern blot analysis indicates that the IGF-BP (BP-25) cDNA hybridizes to a single message transcript of approximately 1.5 kb and is highly tissue specific (5,13). IGF-BP transcripts have been identified in fetal liver but not in other fetal tissues and message levels for IGF-BP are 5 to 10 times greater in fetal than in adult livers (6). In addition, the presence of IGF-BP mRNA in secretory phase endometrium and term decidua, but not in proliferative phase endometrium in both the baboon and the human, suggest that its transcription may be stimulated by progesterone (13,18,41).

Regulation of IGF-BP Synthesis

Before considering the evidence for the hormonal regulation of the synthesis and secretion of IGF-BP, it is relevant to note that only one gene copy appears to exist for this protein and that analysis of the putative promotor region has not revealed any elements which bind specific transcription factors. The presence of a TATA box and CpG island are characteristic of tissue-specific regulated genes. However, the promotor elements required for liver-specific expression were not identified (6,42).

In the non-pregnant adult the identification of specific mRNA in the liver suggests that this organ is the source of the serum protein (BP-53). A large diurnal variation is found with a nocturnal peak, and fasting during the following day results in the maintenance of the nocturnal levels (43). Metabolic studies suggest that serum levels increase during insulin-induced hypoglycemia and fasting, and decrease during a glucose load (43). Other studies have implicated insulin in its regulation (44). This non- pregnant source appears to be regulated by factors primarily associated with metabolism with little involvement of hormonal stimuli derived from the pituitary, thyroid or adrenal cortex (43).

Even though the endometrium during the menstrual cycle does not appear to make any impact upon systemic levels of the BP (45), the amount of protein synthesized by the endometrium during the menstrual cycle is modulated by the gonadal steroids, with the highest levels being observed during the luteal phase (10,13). Progesterone has been reported to increase IGF-BP production by human endometrial explants (46), which had presumably been exposed to endogenous estradiol. Rutanen et al. (46), using polyclonal antibodies reported that this IGF-BP was localized to the glandular epithelium whereas Waites et al. (12) using monoclonal antibodies observed primarily stromal staining. The discrepancies in these observations still need to be resolved. In the baboon, where IGF-BP production is solely localized to the glandular epithelium during the menstrual cycle, progesterone administered to ovariectomized, estradiol-primed animals induces synthesis of IGF-BP in these cells (13). This is consistent with the assumption that synthesis in this cell type is dependent upon progesterone receptor-mediated regulation. In the human during the late luteal phase, IGF-BP production is also induced in populations of stromal cells associated with areas destined to undergo pre-decidualization (see earlier section). Production by these cells may also be dependent upon the same hormonal stimuli that controls stromal differentiation during the menstrual cycle. In endometrium at ectopic sites, ie. endometriosis, IGF-BP production is synchronized with its production in the intra-uterine endometrium (47). Prolactin (PRL) is also a product of the stroma during the late luteal phase (48) and PRL production by endometrial stromal cells is increased by progestins under conditions where decidual cell differentiation is suggested to occur (49,50). Production of PRL and IGF-BP in the human occurs when ovarian steroid levels are decreasing during the late luteal phase. Therefore, whether the synthesis of IGF-BP is a direct effect of progesterone on the stromal cell, or dependent upon the differentiation of the stromal cell into the decidual cell remains unknown. However, during pregnancy decidual IGF-BP production is quantitatively far higher than during the menstrual cycle.

The inability to detect IGF-BP in decidual cells associated with ectopic tubal pregnancy (S.C. Bell and G.T. Waites, unpublished observations) suggests that the production of IGF-BP is not linked with a histologically defined decidual cell per se, but that other regulatory factor(s) must act upon these cells. Recently, employing in vitro culture of endometrial stromal cells, relaxin has been identified as a factor which induces production of IGF-BP such that it represents the major soluble secretory product of cultures and is produced at rates analogous to decidualized endometrial explants (S.C. Bell and L. Tseng, unpublished observations). It is interesting to note that relaxin also enhances the progestin-response of primary stromal cell cultures in terms of aromatase (51) and PRL (49) production. This suggests that the regulation of the products characteristic of the secretory phenotype of the decidual cell may be identical and reflect a specific "decidual cell" response to regulators and the subsequent coordinate expression of these products. It is therefore interesting to note that the mechanism involved in regulation of PRL production from the pituitary are not observed in decidual tissue. It has been suggested that different tissue specific regulatory mechanisms operate in the pituitary and decidual cell for PRL (52). Studies on PRL production from the decidua have implicated a role for exogenous IGF-I (53) and also a trophoblast derived factor (54). This adds further evidence for potential endocrine and paracrine regulation exerted upon the decidual cells.

Function of IGF-BP

The function of decidual IGF-BP, like other IGF-BP's, must be secondary to the function of IGFs, by either modulating their bioavailability to the specific IGF receptors (15) and/or possibly the insulin receptor, or, if the possession of the RGD tripeptide reflects the existence of an 'integrin-type' receptor for this BP, mediating their action via IGF-BP-integrin activation. Production of this BP by decidual tissue, as observed in three primate species (human - 19; baboon - 18; rhesus monkey - S.C. Bell and I.A. Maslar, unpublished observations), must reflect some unique requirement in pregnancy in species exhibiting hemochorial placentation and decidualization (55). The function of decidual tissue may therefore be linked with IGF-BP production, its major soluble secretory product, and may underlay the proposed functions of this tissue in hemochorial placentation. These functions could either be systemic, such as an endocrine function which affects the maternal system, or autocrine and/or paracrine, such as the permissive or restrictive control of trophoblast invasion and placental development, modulation of trophoblast activity, and/or providing an immunologically privileged site via modulation of migration and/or activation of cells of the immuno-inflammatory system (56-58). Basic to these considerations is the report that two forms of the BP, isolated from amniotic fluid on the basis of their charge, exert either inhibitory or stimulatory effects upon IGF-I action upon cells in vitro (32,40).

Systemic action: Since the evidence is consistent with the view that the decidual production of IGF-BP results in the increase in serum BP levels in pregnancy, the increase could fulfill a systemic role related to the putative role of the serum IGF-BP in the non-pregnant adult (see 43). In a systemic role the serum IGF-BP, being substantially unoccupied and able to gain access to the extravascular space, could affect the local bioavailability of IGF's. In non-pregnant adults, Yeoh & Baxter (43) have suggested that the fluctuations in IGF-BP levels may be involved in controlling the relative contribution of IGF's to the action of insulin in the regulation of glucose homeostasis. Thus high IGF-BP levels during hypoglycemia could prevent IGF's acting via insulin receptors to further contribute to hypoglycemia. Conversely in the fed state, the associated lower IGF-BP levels would enhance the bioavailability of IGF's and contribute to insulin-action. It is also possible that such a relationship would ensure that in peripheral tissues IGF action upon cells would be linked to insulin action and enhanced glucose uptake by cells. Therefore, if IGF-BP does contribute to glucose homeostasis in the adult, the decidual source could contribute to the observed alterations in glucose homeostasis associated with pregnancy.

Autocrine action: IGF-BP could regulate IGF-I action upon endometrial/decidual cells, whether IGF's are produced endogenously or exogenously. The endometrium undergoes dramatic growth and differentiation during the first trimester, and if IGF's are required for these processes, the local production of IGF-BP in its stimulatory form could provide a mechanism to enhance only locally, the growth promoting effects of IGF. In decidual cell cultures PRL production is enhanced by exogenous IGF-I (53) and therefore the production of IGF-BP and the bioactive concentration of IGF-I could be involved in autocrine stimulation of PRL production.

Paracrine action: Potential paracrine functions of this IGF-BP may be of great significance when the cellular localization of synthesis and secretion is considered with reference to the behaviour of the embryonic trophoblast during implantation and placental development in the primate. Although decidualization has been proposed to be involved in the control of early trophoblast invasion and growth in hemochorial placental development (see 55), this thesis has been rejected by some authors since histologically-defined decidual cells have not been identified in non-human primates (27). However, at the electron-microscope level 'hypertrophied' stromal cells of the gestational endometrium of these species have been considered to represent 'decidual cells' (59), and this is supported by the common property of the high secretory rates of production of IGF-BP (18; S.C. Bell and I.A. Maslar, unpublished observations). IGF-BP could be involved in inhibiting IGF action upon trophoblast cells, particularly those that invade into the decidualized endometrium as opposed to those that gain assess to blood vessels. However, both stimulatory and inhibitory forms of the IGF-BP could be produced by decidual cells. The presence of the appropriate IGF-BP may vary according to location, and the trophoblast itself could regulate production of decidual IGF-BP such that both inhibitory and promoting effects upon trophoblast activity could occur at different sites. Of great interest is the localization of IGF-BP synthesis to the secretory glandular epithelium at the anticipated time of implantation in both the baboon and human, which is of greater intensity in the former species. A question to be addressed is whether this reflects the mode of implantation in species where the process is superficial as in the baboon or invasive as in the human (30,60). In the sheep, where there is no stromal invasion by the trophoblast, a protein exhibiting immunoreactivity with monoclonal reagents against the human IGF-BP has been localized to the luminal epithelium during the peri-implantation period in pregnant uteri (G.T. Waites, A.T. Whyte and S.C. Bell, unpublished observations). Perhaps this epithelial protein is involved in the transport of IGF's to the uterine lumen thus providing exogenous IGF to the trophoblast. To address these questions basic information concerning the expression and distribution of IGF's, their receptors and BP's in various cellular populations during implantation and pregnancy must be obtained. Studies of this nature are presently underway in our laboratories.

REFERENCES

1. Fazleabas, AT, Verhage, HG (1987). Synthesis and release of polypeptides by the baboon (Papio anubis) uterine endometrium in culture. Biol Reprod 37:979.
2. Fazleabas, AT, Miller, JB, Verhage, HG (1988). Synthesis and release of estrogen and progesterone dependent proteins by the baboon (Papio anubis) uterine endometrium. Biol Reprod 39:729.
3. Bell, SC, James, RFL, Jackson, JA, Patel, SR, Waites, GT, Walczak, K (1989). Monoclonal antibodies to human secretory "pregnancy-associated endometrial α_1-globulin", an insulin-like growth factor binding protein: characterization and use in radioimmunoassay, western blots and immunohistochemistry. Am J Reprod Immunol Microbiol (In Press).
4. Waites, GT, Bell, SC (1989). Immunohistological localization of human pregnancy-associated endometrial α_2-globulin, a glycosylated β-lactoglobulin homologue, in the decidua and placenta during pregnancy. J Reprod Fertil 87:291.
5. Brewer, MT, Stitler, GL, Squires, CH, Thompson, RC, Busby, WH, Clemmons, DR (1988). Cloning, characterization and expression of a human insulin-like growth factor binding protein. Biochem Biophys Res Comm 152:1289.
6. Brinkman, A, Groffen, CAH, Kortleve, DJ, Drop, SLS (1988). Organization of the gene encoding the insulin-like growth factor binding protein IBP-1. Biochem Biophys Res Commun 157:898.
7. Lee, Y-L, Hintz, RL, James, PM, Lee, PDK, Shively, JE, Powell, DR (1988). Insulin-like growth factor (IGF) binding protein complementary DNA from human HEP G2 hepatoma cells: predicted protein sequence suggests an IGF binding domain different from that of the IGF-I and IGF-II receptors. Mol Endocrinol 2:404.
8. Julkunen, M, Seppala, M, Janne, OA (1988). Complete amino acid sequence of human placental protein 14: a progesterone-regulated uterine protein homologous to β-lactoglobulins. Proc Natl Acad Sci (USA) 85:8845.

9. Bell, SC (1988). Secretory endometrial/decidual proteins and their function in early pregnancy. J Reprod Fertil (Supp) 36:109.
10. Bell, SC, Patel, SR, Kirwan, PH, Drife, JO, (1986). Protein synthesis and secretion by the human endometrium during the menstrual cycle and the effect of progesterone in vitro. J Reprod Fert 77:221.
11. Bell, SC, Bohn, H (1986). Immunochemical and biochemical relationship between human pregnancy-associated secreted endometrial α_1- and α_2-globulins (α_1- and α_2-PEG) and soluble placental proteins 12 and 14 (PP_{12} and PP_{14}). Placenta 7:283.
12. Waites, GT, James, RFL, Bell, SC (1988). Immunohistological localization of the human endometrial secretory protein "pregnancy-associated endometrial α_1-globulin (α_1-PEG), an insulin-like growth factor binding protein, during the menstrual cycle. J Clin Endocrinol Metab 67:1100.
13. Fazleabas, AT, Jaffe, RC, Verhage, HG, Waites, G, Bell, SC (1989). An insulin-like growth factor binding protein (IGF-BP) in the baboon (Papio anubis) endometrium: synthesis, immunocytochemical localization and hormonal regulation. Endocrinology 124:2321.
14. Furlanetto, RW (1980). The somatomedin C binding protein: evidence for a heterologous subunit structure. J Clin Endocrinol Metab 51:12.
15. Nissley, SP, Rechler, MM (1984). Insulin-like growth factors: biosynthesis, receptors and carrier proteins. In Hoa Li C (ed): "Hormonal Proteins and Peptides," New York: Academic Press Inc., p 127.
16. Devroede, MA, Tseng, LY-H, Katsoyannis, PG, Nissley, SP, Rechler, MM (1986). Modulation of insulin-like growth factor I binding to human fibroblast monolayer cultures by insulin-like growth factor carrier proteins released to the incubation media. J Clin Invest 77:602.
17. Pavoa, G, Isaksson, M, Jonvall, H, Hall, K (1985). The somatomedin-binding protein isolated from a human hepatoma cell line is identical to human amniotic fluid somatomedin-binding protein. Biochem Biophys Res Comm 128:1071.
18. Fazleabas, AT, Verhage, HG, Waites, G, Bell, SC (1989). Characterization of an insulin-like growth factor binding protein (IGF-BP), analogous to human pregnancy-associated secreted endometrial

α_1-globulin (α_1- PEG) in decidua of the baboon (Papio anubis) placenta. Biol Reprod 40:973.
19. Bell, SC, Patel, SR, Jackson, JA, Waites, GT (1988). Major secretory protein of human decidualized endometrium in pregnancy is an insulin-like growth factor binding protein. J Endocrinol 118:317.
20. Koistenen, R, Kalkkinen, N, Huhtala, M-L, Seppala, M, Bohn, H, Rutanen, E-M (1986) Placental protein 12 is a decidual protein that binds somatomedin and has an identical N-terminal amino acid sequence with somatomadin-binding protein from human amniotic fluid. Endocrinology 118:1375.
21. Ooi, GT, Herington, AC (1988). The biological and structural characterization of specific serum binding proteins for the insulin-like growth factors. J Endocrinol 118:7.
22. Fant, M, Munro, H, Moses, A (1986). An autocrine/paracrine role for insulin-like growth factors in the regulation of placental growth. J Clin Endocrinol Metab 63:499.
23. D'Ercole, AJ (1987). Somatomedins/insulin-like growth factors and fetal growth. J Develop Physiol 9:481.
24. Froesch, ER, Schmid, C, Schwander, J, Zapf, JA (1985). Actions of insulin-like growth factors. Ann Rev Physiol 47:443.
25. Knauer, D, Smith G (1980). Inhibition of biological activity of multiplication-stimulating activity by binding to its carrier protein. Proc Natl Acad Sci (USA) 77:7252.
26. Rivtos, O, Ranta, T, Jalkanen, J, Suikkari, A-M, Voutilainen, R, Bohn H, Rutanen, E-M (1988). Insulin-like growth factor (IGF) binding protein from human decidua inhibits the binding and biological action of IGF-1 in cultured choriocarcinoma cells. Endocrinology 122:2150.
27. Ramsey, EM, Houston, ML, Harris, JW (1976). Interactions of the trophoblast and maternal tissues in three closely related primate species. Am J Obstet Gynecol 124:647.
28. Waites, GT, James RFL, Bell, SC (1989). Human pregnancy-associated endometrial α_1-globulin, an insulin-like growth factor binding protein: immunohistological localization in the decidua and placenta during pregnancy employing monoclonal antibodies. J Endocrinol 120:351.

29. Hearn, JP (1986). The embryo-maternal dialog during early pregnancy in primates. J Reprod Fertil 76:809.
30. Hendrickx, AG, (1971). "Embryology of the Baboon." Chicago: University of Chicago Press.
31. Bell, SC, Smith S (1988). The endometrium as a paracrine organ. In Chamberlain GVP (ed): "Contemporary Obstetrics and Gynecology," London: Butterworths Scientific Ltd, p 273.
32. Elgin, RG, Busby, WH, Clemmons, DR (1987). An insulin-like growth factor binding protein enhances the biologic response to IGF-I. Proc Natl Acad Sci (USA). 84:3254.
33. Blum, WF, Jenne, EW, Reppin, F, Kietzmann, K, Ranke, MB, Bierich, JR (1989). Insulin-like growth factor I (IGF-I)-binding protein complex is a better mitogen than free IGF-I. Endocrinology 125:766.
34. Wood, WI, Cachianes, G, Henzel, WJ, Inslow, GA, Spencer, SA, Helliss, R, Martin, JL, Baxter, RC (1989). Cloning and expression of the growth hormone-dependent insulin-like growth factor binding protein. Mol Endocrinol 2:1176.
35. Huhtala, ML, Koistinen, R, Palomaki, P, Partanen, P, Bohn, H, Seppala, M (1986). Biologically active domain in somatomedin-binding protein. Biochem Biophys Res Comm 141:263.
36. Ebina, Y, Ellis, L, Jarnagin, K, Edery, M, Graf, L, Clauser, E, Ou, JH, Masiarz, F, Kan, YW, Goldfine, ID, Roth, RA, Rutter, WJ (1985). The human insulin receptor cDNA: the structural basis for hormone-activated transmembrane signalling. Cell 40:747.
37. Johnson, D, Lanahan, A, Buck, CR, Sehgal, A, Morgan, C, Mercer, E, Bothwell, M, Chao, M (1986). Expression and structure of the human NGF receptor. Cell 47:545.
38. Ullrich, A, Gray, A, Tam, AW, Yang-Feng, T, Tsubokawa M, Collins, C, Henzel, W, LeBon, T, Kalhuria, S, Chen, E, Jacobs S, Franke, U, Ramachandran J, Fujita-Yamaguchi, Y (1986). Insulin-like growth factor I receptor primary structure: comparison with insulin receptor suggests structural determinants that define functional specificity. EMBO J 5:2503.
39. Ruoslahti, E, Pierschbacher, MD (1986). Arg-Gly-Asp: a versatile cell recognition signal. Cell 44:517.
40. Busby, WH, Klapper, DG, Clemmons, DR (1988). Purification of a 31,000 dalton insulin-like growth

factor binding protein from human amniotic fluid. J Biol Chem 263:14203.
41. Julkunen, M, Koistenen, R, Aalto-Setala, K, Seppala, M, Janne, OA, Konatula, K (1988). Primary structure of human insulin-like growth factor binding protein/placental protein 12 and tissue specific expression of its mRNA. FEBS Letters 236:295.
42. Cubbage, ML, Suwanichkul, A, Powell, DR (1989). Structure of the human chromosomal gene for the 25 kilodalton insulin-like growth factor binding protein. Mol Endocrinol 3:846.
43. Yeoh, S-T, Baxter, RC (1988). Metabolic regulation of the growth hormone independent insulin-like growth factor binding protein in human plasma. Acta Endocrinol 119:465.
44. Suikkari, A-M, Koivisto, VA, Rutanen, E-M, Yki-Jarvinen, H, Karonen, S-L, Seppala, M (1988). Insulin regulates the serum levels of low molecular weight insulin-like growth factor-binding protein. J Clin Endocrinol Metab 66:266.
45. Suikkari, A-M, Rutanen, E-M, Seppala, M (1987). Circulating levels of immunoreactive insulin-like growth factor-binding protein in non-pregnant women. Hum Reprod 6:297.
46. Rutanen, E-M, Koistinen, R, Sjoberg, J, Julkunen, M, Wahlstrom, T, Bohn, H, Seppala, M (1986). Synthesis of placental protein 12 by human endometrium. Endocrinology 118:1067.
47. Waites, GT, James, RFL, Walker, RA, Bell, SC (1989) Human pregnancy-associated endometrial α_1-globulin, a 32 KDa insulin-like growth factor binding protein: immunohistochemical distribution and localization in the adult and fetus. J Endocrinol (In Press).
48. Maslar, IA, Riddick, DH (1979). Prolactin production by human endometrium during the normal menstrual cycle. Am J Obstet Gynecol 135:751.
49. Huang, JR, Tseng, L, Bischof, P, Janne, OA (1987). Regulation of prolactin production by progestin, estrogen and relaxin in human endometrial stromal cells. Endocrinology 121:2011.
50. Tseng, L, Malbon, CC, Lane, B, Kaplan, C, Mazella, J, Dahler, H, Tseng, A (1987). Progestin-dependent effect of forskolin on human endometrial aromatase activity. Hum Reprod 2:317.

51. Tseng, L, Mazella J, Chen, G (1987). Effect of relaxin on aromatase activity in human endometrial stromal cells. Endocrinology 120:2220.
52. Harman, I, Costello, A, Ganong, B, Bell, RM, Handwerger, S (1986). Activation of protein kinase C inhibits synthesis and release of decidual prolactin. Endocrinol Metab 14:172.
53. Trailkill, KM, Golander, A, Underwood, LE, Handwerger, S, (1988). Insulin-like growth factor I stimulates the synthesis and release of prolactin from human decidual cells. Endocrinology 123:2930.
54. Handwerger, S, Barry, S, Markoff, E, Barrett, J, Conn, PM (1983). Stimulation of the synthesis and release of decidual prolactin by a placental polypeptide. Endocrinology 112:1370.
55. Bell, SC (1983). Decidualization: regional differentiation and associated function. Oxf Rev Reprod Biol. 5:220-271.
56. Bell, SC (1985). Comparative aspects of decidualization in rodents and human: cell types, secreted products and associated function. In Edwards RG, Purdy J, Steptoe PC (eds): "Implantation of the Human Embryo," London: Academic Press, p 71.
57. Bell, SC (1989). Decidualization and insulin-like growth factor (IGF) binding protein: implications for its role in stromal cell differentiation and the decidual cell in haemochorial placentation. Hum Reprod 4:125.
58. Bell, SC (1989). Decidualization and relevance to menstruation. In "Contraception and Mechanisms of Endometrial Bleeding," Cambridge University Press, (In Press).
59. Enders, AC, Welsh, AO, Schlafke, S (1985). Implantation in the Rhesus Monkey: endometrial response. Am J Anat 173:147.
60. Enders, AC, Schlafke, S (1986). Implantation in non-human primates and in the human. Comp Prim Biol 3:291.

IN VITRO IMPLANTATION ON POLARIZED UTERINE EPITHELIA[1]

Stanley R. Glasser,[2] Joanne Julian,[2] Joy Mulholland,[2] Shailaja Mani,[2] Daniel D. Carson[3] and Andrew L. Jacobs[3]

[2]Department of Cell Biology, Baylor College of Medicine and [3]Department of Biochemistry and Molecular Biology, M.D. Anderson Cancer Institute Houston, Texas 77030

ABSTRACT Blastocyst attachment is a hormonally regulated specialized function of uterine epithelial (UE) cells. In vivo attachment to the apical UE surface occurs only during a short period following estrogen (E) modulation of the progesterone (P) dominated rat uterus; 24-30 h after E the UE cells become non-receptive to embryo attachment. The temporal, spatial and biochemical constraints which regulate in utero attachment do not apply when the blastocyst is removed to ectopic or in vitro environs. Regardless of hormonal conditions blastocysts, in vitro, attach without discrimination to almost any surface, including UE cells. The inability to modulate in vitro UE receptivity to attachment has been attributed to the failure of UE cell cultures to develop polarity.

Primary cultures have been developed in which UE cells achieve polarity and maintain responsiveness to E. Cultured on EHS matrix-impregnated HA or CM Millipore filters in defined medium containing 2.5 x 10^{-9} M E_2, UE cells proliferate to confluence and exhibit morphological and functional polarity. Polar organization, represented by separate plasma membrane domains, functional tight junctions and trans-epithelial resistance is validated by ultrastructural and immunocytochemical evidence. Coordinated indices

---

This work was supported by NIH grants HD-25189, HD-13663 and HD-07495.

of functional polarity include preferential basal surface uptake of ^{35}S-methionine and apical preference for increased UE secretory activity. Development of polarity was monitored by analysis of protein and glycoconjugate profiles of apical and basolateral secretory compartments. Expression of CAM 105 (apical membrane integrated glycoprotein) and apical secretion of two marker proteins, induced in uteri of immature rats by E_2, confirm the hormonal responsiveness of these polarized UE cells.

In the presence of E_2 95% of the blastocysts transferred directly or after prior culture with polarized UE cells attach within 24-48h and grow out on bare HA filters, EHS-impregnated HA or CM filters or on uterine stromal cell monolayers on EHS covered filters. Only when co-cultured with polarized UE cells in the presence of E_2 do blastocysts fail to attach. They remain viable and continue to develop. That the development of an apical UE surface nonreceptive to blastocyst attachment is an expression of the responsiveness of these cells to E_2 is reinforced by CAM 105 expression and the secretion of uterine E_2 marker proteins. Regulation of blastocyst attachment, a specialized function of the UE cell, is dependent in part on its polarity.

INTRODUCTION

In vitro models of oocyte maturation and embryo development provide reasonably faithful and useful replicas of in vivo development. These reliable, reproducible culture systems that support the progression of in vitro fertilized mouse ova through cleavage, compaction, hatching and attachment allow almost a full menu of morphological, biochemical and molecular biological studies of natural and micromanipulated embryos. These protocols are less reliable when applied to ova and embryos of other species, i.e., rat, rabbit, cow, sheep, human. More than 80% of rat blastocysts recovered from a receptive uterus will attach when transferred to a synchronous recipient. However, rat zygotes have not been reliably or reproducibly cultured past the 2-cell stage .

Culture of bovine zygotes in classic cell culture systems has also been difficult. Of those bovine embryos which become blastocysts only 30% attached (1). Some improvement over in vitro culture has been achieved by incubating zygotes in ligated sheep oviducts until transfer. Thus, it seems reasonable to suggest the three-dimensional, hormonally responsive oviductal and uterine epithelial cell cultures would be required to provide the paracrine factors that could promote in vitro developmental progression of bovine zygotes through blastocysts. Although based, at least in part, on this strategem new, innovative procedures for processing human ova have yet to produce the rate of implantation success expected of them.

Criticism of conventional, static culture systems as models for early mammalian development, particularly implantation related events, has been documented (2). In these cases the failure of epithelial cells to establish the three-dimensional polarity which has been correlated with their ability to express morphological and functional differentiation has been attributed, in part, to the absence from these culture systems of a substratum, either extracellular matrix proteins or a cellular feeder layer (3). Some concern should also be directed to the formulation of culture media, particularly serum-free, completely defined media. Sources of essential fatty acids, glycolipids, high and low density lipoproteins, optimally compatible with the energy requirements of developing embryos, are conventially absent from culture media. What these data suggest is that conventional culture systems do not provide, directly or indirectly, a source of paracrine factors (specific energy sources, growth and other regulatory agents) that are critical to the progressive development of certain mammalian embryos. A recent report on the in vitro development of human embryos underscores the importance of the composition of the culture environment. Human zygotes, resulting from in vitro fertilization (IVF), were cultured in either a modified Earles' balanced salt solution in the presence or absence of a monolayer of fetal bovine uterine fibroblasts. At the time of embryo transfer human zygotes cultured in the presence of fetal uterine fibroblasts experienced less fragmentation and possessed more blastomeres. The incidence of implantation was greater with co-cultured embryos (35%) than with embryos developing in Earles' Balanced salt solution (4). Similar experience has been reported with embryos of other species (1).

The cause and effect relationships which underlie the shift from a nonreceptive uterus to a uterus transiently receptive to blastocyst attachment have been assigned to the resultant responses of the hormonally regulated UE cell (5) manifested by morphological and functional components of its plasma membrane domains. In vivo receptivity is transitional. Should blastocyst attachment fail to occur a second nonreceptive phase follows. This refractory period is different from the pre-implantation interval (5,17). It has been predicted that this transition would be marked by further hormonally regulated shifts in the composition of the apical UE surface which are no longer compatible with cell-cell adhesion (6).

Efforts to define these peri-implantation changes in ovoreceptivity in terms of biochemistry were spurred by episodic reports of stage related shifts in the profile of secretory and cell associated uterine proteins and glycoproteins (7,8). The usefulness of these studies was faulted by the cell heterogeneity of the endometrium.

If the complexities of endometrial-blastocyst interactions are to be resolved it is important to be able to apply recent advances in cell and molecular biology to these analyses. Separation of individual endometrial cell types, and their culture, would provide greater access. Definition of the potential role of each individual cell type will allow pivotal questions regarding their separate contributions to the peri-implantation status of the uterus to be resolved. Used as components of cell-cell recombinants their expression will allow us to identify and define the mechanisms by which instructive and/or modulating signals (endocrine, paracrine, autocrine) serve to stimulate, repress or derepress the regulatory factors which render uterine epithelial cells nonreceptive, receptive or refractory to the blastocyst.

The use of in vitro models to resolve these questions has been unsatisfactory. None of the regulatory programs which governs receptivity in vivo (hormones, time, tissue organization) has any influence in vitro; blastocysts will attach, without qualification, to almost any surface. Culture systems designed to study hormonal responses of isolated UE cells have yielded unsatisfying results whether assessed in terms of cell proliferation or differentiation of specialized functions including attachment. Nevertheless, individual UE and US cell cultures are theoretically preferred to endometrial cell cultures because they resolve the issue of cell heterogeneity and reduce the competition between UE and US cells. UE cell cultures may be reasonably adequate for certain short term studies. When used in protocols longer than 48-60h they can be the source of much disinformation. Assuming satisfactory separation of endometrial cells into separate homogeneous populations of viable UE and US cells (9,10) most culture systems do not provide an environment that enables UE cells to attach, proliferate to confluence (Fig. 1) and establish polarity. The validity of in vitro implantation models has been justifiably censured because they do not provide a three-dimensional, polarized environment (2).

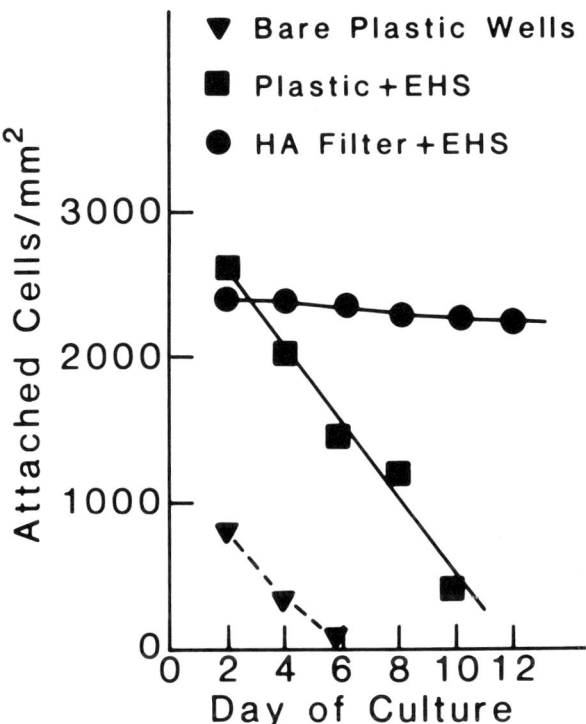

Figure 1. Effect of culture surface on maintenance of attached UE cells. (●) Cells cultured on EHS matrix-coated filters reach confluence between 48-72h, depending on seeding density, and maintain their confluent density. (■) Cells cultured on matrix-coated wells also become confluent between 48-72h but begin to detach after achieving confluence and are gone by days 12-14. (▼) Cells cultured on bare tissue culture plastic never achieve confluence. Those cells which attach do so transiently; they are gone by day 6.

Epithelial Cell Polarity

The ability of absorptive, secretory and transporting epithelial cells, and certain epithelial cell lines, to express their special functions, depends on polarization of their plasma membranes. Polarity can be defined in terms of morphological, compositional and functional destruction of the cells apical vs. basolateral domains of considerable importance to the establishment and maintenance of polarized functions in the integrity of intercellular functions and environmental signals, particularly those presented via the substratum (3,11). These principles derive mostly from studies of established cell lines. Few studies have attempted to define epithelial phenotypic expression in terms of polarity of primary epithelial cells; none have been reported with UE cells.

Establishing Polarized Cultures of Uterine Epithelial Cells

Renal, intestinal or hepatic primary epithelia attach efficiently (~75%) within 24h. These cells proliferate to confluence within 48h and polarize (12). UE cells are not as efficient. Only 20% attach within 24h when cultured on plastic in the presence of serum; ~60% when cultured in the absence of serum. In neither case do the cells proliferate at a rate to achieve confluence and cells detach so none remain after 6d (Fig. 1). While culture on extracellular matrix will enhance attachment (13) UE cells appear to have special requirements. Thus 80% of UE cells attach to collagen I after 24h culture in complete medium. However attachment is transient; UE cells do not proliferate to confluence and few cells remain after 12d. Before cells can establish polarity they must become confluent.

Extrapolation of data from established epithelial cell lines and some primary epithelial cells (12) would predict that cells which do not achieve confluency, that do not polarize in culture, would fail to express their complement of specialized functions. Extended to UE cells this would mean that the cells would not retain their ability to respond to P and/or E. Although surviving cells may continue to function metabolically their secretions may be more representative of homeostasis rather than the sequence of hormonally regulated stage-specific responses. Nonpolarized UE cells are not able to morphologically and functionally transpose from nonreceptive to receptive to refractory.

To develop a reliable in vitro model of implantation we introduced two modifications (14). UE cells were obtained from endocrine naive (immature) UE rats to minimize variability associated with UE cells from sexually mature (intact, castrated) or pregnant rats. A basement membrane-like matrix derived from Engelbreth-Holm-Swarm (EHS) tumor cells was used in place of collagen I. The composition of EHS matrix is laminin (84%), entactin (11%), collagen IV (4%), and heparin sulfate proteoglycan (1%). UE cells cultured in basal medium on EHS matrix will, depending on seeding density, attach and proliferate to confluence within 48-72h. Under these conditions the cells begin to detach so, even on EHS matrix, UE cells are effectively gone by d.12 (Fig. 1). Confluence is not synonymous with polarity.

Since it had been established that, in addition to specific matrix requirements, a permeable culture surface was also important to the function of epithelial cells (15) UE cells were cultured on EHS impregnated permeable HA filters. UE cells on these components of the bicameral Millicell systems (Millipore, Bedford MA) achieved and maintained confluent densities for the duration of these studies (Fig. 1). The ability of these primary UE cultures to establish polar organization, separate apical vs. basolateral plasma membrane domains and form tight junctions was validated by immunocytochemical and ultrastructural evidence (14). Coordinately with the development of polarity in post-confluent cultures transepithelial resistance increased from $<20\Omega/cm^2$ to $>300\Omega/cm^2$. Freshly cultured UE cells appear flattened but display interdigitated lateral junctions and rudimentary microvilli. Confluent UE cells are tightly adherent to each other (zonae adherens, desmosomes) and display elongated microvilli. Polarized UE cells express their signature cytokeratins (CK 7,8,18 and 19). The absence of cytokeratin negative cells proves the effectiveness of the cell separation method (9,10). The progress of plasma membrane reorganization during the transition from preconfluent to postconfluent UE cells may be tracked by expression of uvomorulia (E-Cadherin). The appearance of this plasma membrane protein which follows the formation of tight junctions is restricted to the basolateral surface of polarized UE cells (14).

Indices of functional polarity develop as correlates of morphological polarity (14,16). Reliable assays of the differences between apical vs. basolateral plasma membrane domains include (a) the preferential uptake of ^{35}S-methionine from the basal surface (Fig. 2A), (b) a significant increase in total secretory activity of the UE

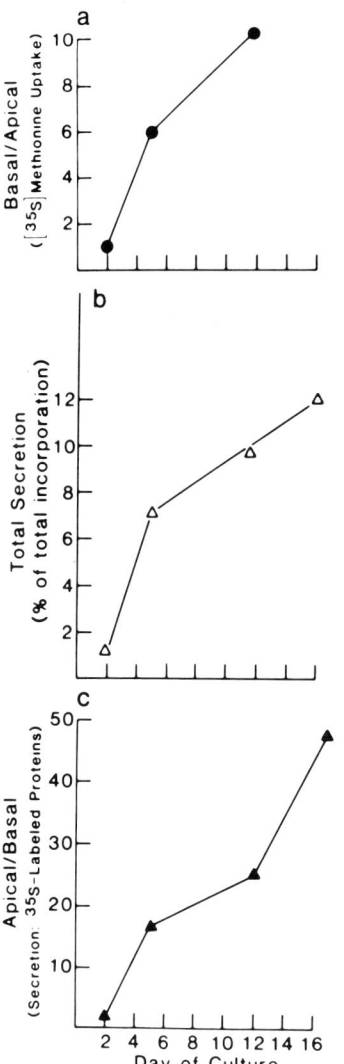

Figure 2. (a) Development of the basolateral preference for uptake of $[^{35}S]$-methionine in UE cultures on matrix-coated filters. Uptake (10 min) was measured from either the apical or basal side of the filter. Data is expressed as the ratio of basal/apical TCA-precipitable cpm taken up by filter cultured cells on day 2 (pre-confluent), day 5 (early post-confluent) and day 12 (late post-confluent). **(b)** Development of increased secretory activity in post-confluent UE cells cultured on matrix-coated filters. Total secretion (TCA-precipitable cpm in apical plus basal secretions) expressed as a percentage of total TCA-precipitable cpm incorporated (8 hrs, $[^{35}S]$-methionine) into UE cells plus their secretions. **(c).** Development of apical preference for secretion. UE cells cultured on matrix-coated filters were labeled with $[^{35}S]$-methionine for 8h. The data represent the ratio of total TCA-precipitable cpm in the apical relative to the precipitable cpm in the basal secretory compartment (14).

cell (Fig. 2b) and (c) a marked preference for apical secretion which increases with time in culture (Fig. 2c).

Differences in profiles of proteins and glycoproteins of the apical and basolateral secretory compartments proved to be dependable measures of the development of functional

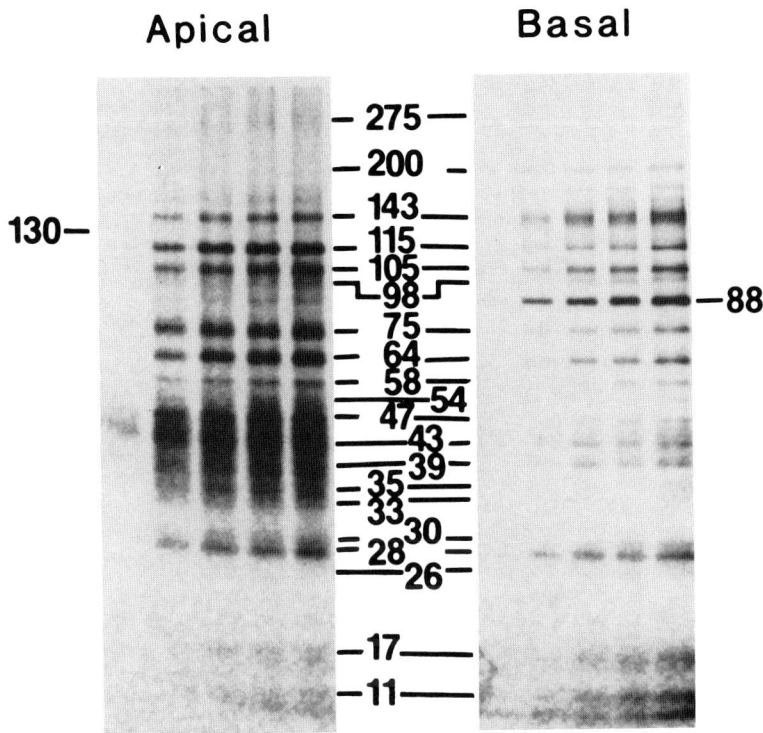

Figure 3. Apical and basal secretory products were collected at 2,4,6,8 and 10h of incubation with [^{35}S]-methionine (from left to right) from UE cells cultured on matrix-coated filters for 11 days. Autoradiographs of SDS-gels were exposed for 18h (apical) and 7 days (basal). Molecular mass (X10^{-3}) is designated by number. Components common to both secretory compartments are indicated between the two gels while components which are either uniquely (130 KD) or preferentially (88 KD) secreted to a particular compartment are indicated at the sides. Secretion in both compartments was linear with time through the 10th hour of incubation. The average apical/basal ratio for all timepoints was 26 (14).

polarity. Polarized UE cell cultures could be identified by a 130 Kd protein, unique to the apical compartment, and an 88 Kd protein preferentially secreted to the basal compartment (Fig. 3) (14). With increased time in culture the secretion of those proteins and glycoproteins common to both secretory compartments (Fig. 3) shifted thereby contributing to the increase in apical secretory activity (Fig. 2c). Proteoglycans, i.e., keratin sulfate proteoglycan (KSPG) and heparan sulfate containing proteoglycan (HSPG), were also major products synthesized and secreted by polarized UE cells (16). The apical secretion of KSPG increased as a function of developing polarity. On the other hand, HSPG secretion, also preferentially apical, was not coordinated with polarity. This pattern of proteoglycan secretion mimicked that produced by uterine strips.

Amongst the notable characteristics of this culture system, in addition to its competence to establish morphological and functional polarity, is the maintenance of hormone responsiveness. Thus, the in vitro development of the secretory phenotype in polarized UE cells recommends the use of this system to resolve the mechanisms which govern the transitions of the peri-implantation uterus. While the degree of relationship between hormonal responsiveness and polarity dependent expression of UE special functions remains to de defined the progression from nonreceptivity to receptivity to refractoriness has been related to changes in the synthetic and/or secretory activity of UE cells (17,18).

Redefinition of the relationship between the basal UE surface and the basal lamina and developmental changes in the extracellular matrix of decidualized stromal cells could modify the reciprocal relationships between UE and US cells. Therefore ability to monitor alterations at the basal surface of the UE cell is an unique property of this culture system that is intrinsic to understanding the regulatory biology of the UE cell. Basal secretions could include transduced signals, i.e., histamine. estrogens, prostaglandins, that presumptively could serve to initiate stromal cell differentiation. Thus it is significant that, for the first time the preferential secretion of $PGF_2\alpha$, via the basal surface of the UE cell, has been demonstrated (A. Jacobs, personal communication). Upon attaining confluency UE cells isolated from either mouse or rat and cultured on matrix impregnated Millicell HA filters, preferentially secreted $PGF_2\alpha$ to the basal secretory compartment (basal/apical=2.5-4.1). Neither total prostaglandin secretion nor the basal/apical ratio was disrupted by incubating polarized UE cells with indomethacin (15 µg/ml). $PGF_2\alpha$ secretion was energy dependent (sensitive to 10mM sodium azide) but was not coupled with protein synthesis (cycloheximide (6µg/ml). The correlation between UE cell polarity and the vectorial secretion of $PGF_2\alpha$ was established by studies with EGTA (8mM) which reduced the basal/apical differential in $PGF_2\alpha$ secretion to near unity. Contrary to their behavior in other culture systems polarized UE cells are not only hormonally responsive but their response to E is similar to that produced by the UE cell in situ. Secreted proteins (19) and glycoproteins (20) have been identified as markers of E action produced by uterine strips of hormonally stimulated rats. Identical subunits (115 Kd, 64 Kd) of complement protein C_3 (21) have been identified in the apical secretion of polarized UE cells (14). Studies of proteoglycan synthesis and secretion by polarized UE cells demonstrated that 80-90% of KSPG and HSPG were preferentially secreted to the apical surface (16). These results are consistent with earlier studies which showed E stimulated HSPG expression at the luminal surface of whole uteri (22,23).

Receptivity and the Adhesive Stage of Implantation

The hormonally regulated apical disposition of proteins and glycoproteins has been implicated as the operative (paracrine?) factor in the intrauterine shift from a nonreceptive to receptive UE cell. Since both proteoglycans and those proteins which specifically bind proteoglycans are expressed on both UE and blastocyst surfaces it is possible that these are the biochemical components of the apposition/adhesive stages of implantation. This type of cell-cell adhesion system has been described for other cell systems.

Attempts to simulate this type of cell-cell adhesion system to enlarge our understanding of the regulatory biology of the UE cell, particularly its role in blastocyst attachment, have been remarkably nonproductive. None of the temporal or hormonal constraints which govern in vivo blastocyst attachment appear to exercise any influence on the in vitro behavior of blastocysts. As long as they are exposed to optimal nutritive conditions (at least 24h) more than 80% of blastocysts recovered from uteri of rats, mice and hamsters will become adhesive, attach and undergo trophoblast outgrowth. This absence of specificity can be assigned to the failure of in vitro culture systems to support UE cell polarity (2,24). Development of UE cell polarity correlated with the maintenance of hormonal responsiveness and thereby the expression of specialized, possibly paracrine, UE cell functions suggests that this model could be productively used as an experimental in vitro implantation model.

Blastocysts were recovered from the uteri of pregnant rats on the day of implantation (d.4) and transferred to various culture surfaces in media selected to provide optimally favorable nutritional and attachment conditions. Complete media contained E_2 at a concentration of 2.5×10^{-9}M (stimulates expression of marker proteins in immature rats). All other steroids were below their effective titers. Within 48h blastocysts attachment (90-95%) occurred and was followed by trophoblast outgrowth on bare plastic wells, bare HA filters and filters covered with EHS matrix (Table 1). There was no attachment to bare CM filters but if they were first covered with EHS attachment was >90%. However if a polarized UE culture was developed on EHS covered CM filters then no blastocysts attached over the 120h study period (Table 1).

To determine if blastocysts cultured on bare CM filters were adversely affected by the filter they were transferred, after 48h culture, to CM filters covered with EHS. 80-85% of these blastocysts attached within 24h (Tab). However if the transfer was made from bare CM filters to polarized UE cultures none of the blastocysts ever attached (Table 1).

The same questions about implantation competency were asked about blastocysts which failed to attach to polarized UE cultures. Of those blastocysts transferred to EHS covered CM filters (no cells) after varying intervals of culture in the presence of UE cells (24-72h) over 80% attached within 24h of transfer. Additional groups were transferred at the same time from polarized UE cultures to monolayer cultures of stromal cells on EHS covered CM filters. In every case at least 95% of the blastocysts attached within 24h after transfer from UE to US cells (Table 1).

These experiments secure the observation that blastocysts cultured on polarized UE in the presence of E_2 cells fail to attach. They retain their ability to do so, even in the presence of E_2, when transferred to another culture environment. Thus, for the first time, we have been able to impose experimental constraints on the in vitro attachment of viable adhesive blastocysts. E_2 renders polarized UE cells but not US cells nonreceptive. Whatever effect E_2 has on UE cells it does not have on US or trophectoderm cells. That polarized UE cells respond directly to E_2 is analogous to in vivo conditions that do not allow attachment. Thus nonreceptivity may be considered an E_2 regulated expression of a specialized function of polarized UE cells.

TABLE 1
IN VITRO ATTACHMENT OF RAT BLASTOCYSTS TO VARIOUS SUBSTRATES

Substratum	% Attachment at			
	48h	72h	90h	120h
(A) Bare plastic wells	90	94	94	94
Bare CM filters	0	0	0	0
CM + EHS [a]	92	94	96	98
CM + EHS + UE[b]	0	0	0	0
(B) Transfer from bare CM filters to				
CM + EHS (48h)	–	83	–	96
CM + EHS + UE (48h)	–	0	0	0
(C) Transfer from CM + EHS + UE to				
CM + EHS (48h)	–	82	88	90
CM + EHS + US[c] (48h)	–	95	95	98

[a] Engelbreth-Holm-Swarm tumor matrix
[b] Primary cultures of immature uterine epithelial cells
[c] Primary cultures of immature uterine stromal cells

These data (14,16) indicate the response of immature UE cells to E_2 is direct and not solely dependent on signals from hormonally responsive US cells. Since the culture system allows access to the basal surface of the cell it is now possible to include analyses of UE→US interactions. Coupled with the definition of behavior of isolated UE cells alone cell-cell recombinations can be selectively constructed to extend our understanding of the role of attachment and post-attachment differentiation of both endometrial cell types in initiation of placentation.

REFERENCES

1. Lindenberg S, Hyttel P, Sjgren A, Greve T (1989). A

comparative study of attachment of human, bovine and mouse blastocysts to uterine epithelial monolayer. Hum Reprod 4:446.
2. Enders AC, Chavez DJ, Schlafke S (1981). Comparisons of implantation in utero and in vitro. In Glasser SR, Bullock DW (eds) "Cellular and Molecular Aspects of Implantation" New York: Plenum Press, p 365.
3. Bissell MJ, Hall HC, Parry G (1982). How does the extracellular matrix direct gene expression? J Theor Biol 99:31.
4. Wiemer KE, Cohen J, Amborski GF, Wright G, Wiker S, Munyakazi L, Godke RA (1989). In vitro development and implantation of human embryos following culture on fetal bovine uterine fibroblast cells. Hum Reprod 4:595.
5. Psychoyos A (1973). Endocrine control of egg implantation. In Greep RO, Astwood EB (eds) "Handbook of Physiology, Sect 7, Endocrinology, Pt II" Washington DC: Am Physiol Soc, p 187.
6. Glasser SR, Julian JA, Mani SK, Mulholland J, Munir MI, Lampelo S, Soares MF (1989). Blastocyst-endometrial relationships: reciprocal interactions between uterine epithelial and stromal cells and blastocysts. Troph Res 5 (in press).
7. Anderson TL, Zullo F, Coddington CC, Hodgen GD (1989). Both saccharide-containing and saccharide-binding membrane proteins are expressed in human endometrium during the window of uterine receptivity to implantation, Abst 485, Soc Gyn Invest, 36th Mtg, San Diego CA.
8. Surani MAH (1975). Hormonal regulation of proteins in the uterine secretions of ovariectomized rats and the implications for implantation and embryonic diapause. J Reprod Fert 43:411.
9. McCormack SA, Glasser SR (1980). Differential response of individual uterine cell types from immature rats treated with estradiol. Endocrinology 106:1634.
10. Glasser SR, Julian JA (1986). Intermediate filament protein as a marker for uterine stromal cell decidualization. Biol Reprod 35:463.
11. Emmerman JT, Burwen SR, Pitelka DR (1979). Substrate properties influencing ultrastructural differentiation of mammary epithelial cells in culture. Tissue and Cell 11:109.
12. Simons K, Fuller SD (1985). Cell surface polarity in epithelia. Annu Rev Cell Biol 1:243.
13. Madden ME, Sarras MP (1985). Development of an apical plasma membrane domain and tight junctions during

histogenesis of the mammalian pancreas. Dev Biol 112:427.
14. Glasser SR, Julian JA, Decker GL, Tang J-Y, Carson DD (1988). Development of morphological and functional polarity in primary cultures of immature rat uterine epithelial cells. J Cell Biol 107:2409.
15. Matlin KS (1986). The sorting of proteins to the plasma membranes in epithelial cells. J Cell Biol 103:2565.
16. Carson DD, Tang J-Y, Julian JA, Glasser SR (1988). Vectorial secretion of proteoglycans by polarized rat uterine epithelial cells. J Cell Biol 107:2425.
17. Mulholland J, Leroy F (1989). Protein and mRNA synthesis in the peri-implantation rat endometrium. Serono Symposium on Blastocyst Implantation (in press).
18. Roberts RM, Bazer FW (1988). The functions of uterine secretions. J Reprod Fert 82:875.
19. Wheeler C, Komm BS, Lyttle CR (1987). Estrogen regulation of protein synthesis in the immature rat uterus: The effects of progesterone during in vitro incubation. Endocrinology 120:910.
20. Takeda A, Takahashi N, Shimazu S (1988). Identification and characterization of an estrogen inducible glycoprotein (USP-1) synthesized and secreted by rat uterine epithelial cells. Endocrinology 122:105.
21. Lyttle CR, Sundstrom SA, Ponce-de-Leon H, Komm BS (1989). Identification of an estradiol induced rat uterine secretory protein as complement component C_3. Abst 725, Endocrine Society Mtg, Seattle WA.
22. Tang J-Y, Julian J, Glasser SR, Carson DD (1987). Heparin sulfate proteoglycan synthesis and metabolism by mouse uterine epithelial cells cultured in vitro. J Biol Chem 262:12832.
23. Morris JE, Potter SW, Gaza-Bulesco G (1988). Estradiol induces an accumulation of free heparin sulfate glycosaminoglycan chains in uterine epithelium. Endocrinology 122:242.
24. Sengupta J, Given RL, Carey JB and Weitlauf HM (1986). Primary culture of mouse endometrium on floating collagen gels: a potential model for implantation. Annals NY Acad Sci 476:75.

Index

Aborted pregnancies in mice, in epidermal growth factor deficiency, 126
β-Actin in preimplantation mouse embryos, 15
Actin mRNA in mouse oocytes or preimplantation embryos, 35
ADP content in mouse oocytes, 68
Aerobic glycolysis in preimplantation mouse embryos, 72–75
Albumin, bovine serum, in culture medium, and hamster embryo development, 82
Alkaline phosphatase gene detection in early mouse embryos, 37–40
α-Amanitin, and protein synthesis in embryos, 7, 32
Amino acids
   and energy metabolism in preimplantation embryo, 74
   in epidermal growth factor, 126
   and hamster embryo development, 81, 92
   in insulin, 110
   in insulin receptors, 116
Amniotic fluid insulin-like growth factor binding protein in primates, 138, 139, 141, 145
AMP content in mouse oocytes, 68
Amphibians. *See* Frog embryos; Xenopus embryos
Angiogenesis, growth factors affecting, 20
Ascites tumor cells, metabolism in, 84
Aspartic acid, and hamster embryo development, 83
ATP content in mouse oocytes, 68
Autoradiography, and detection of insulin receptors in mouse embryos, 110, 117

Baboons, insulin-like growth factor binding protein and pregnancy in, 137–147
Blastocoel formation in mouse
   gene expression in, 97–105
   and glucose utilization, 73, 75
Blastocyst
   factors affecting development in hamsters, 81–93
   interactions with endometrium, 156
   in vitro attachment on polarized uterine epithelia, 153–165
   juxtacoelic distribution of Na^+,K^+-ATPase catalytic subunits, 97–105
Blastomere swelling in 2-cell hamster embryos, 82, 93
Blocks to development
   in bovine embryos, 4, 8
   in hamster embryos, glucose and phosphate affecting, 79–94
Bovine blastocysts, attachment to uterine epithelia, 154–155
Bovine embryo development
   blockage at 8-16 cell stage, 4, 8
   cell cycle 1 in, 5–6
      DNA analysis in, 5–6
   cell cycle 2 in, 6
   cell cycle 3 in, 6–7
      asynchronous cleavage in, 6–7
      protein synthesis in, 7
      sensitivity to culture conditions in, 7
   cell cycle 4 in, 7–8
      protein synthesis in, 8
   and co-culture of oocytes
      with granulosa cells or cumulus cells, 4
      with oviduct epithelial cells, 4, 8
   and cumulus cell expansion, 3
   cytoplasmic maturation in, 3–5
   energy metabolism in, 70
   follicle size affecting, 3–4
   oocyte maturation in, 2–3
   sperm concentrations affecting, 4
   and synchronous sperm penetration and oocyte activation, 5
Bovine serum, fetal, and responsiveness of trophoblast cells, 48–49
Bovine serum albumin in culture medium, and hamster embryo development, 82
Bovine uterine fibroblasts, fetal, cultured with human embryos, 155

Index

Calcium levels, intracellular, affected by epidermal growth factor, 126
CAM 105 expression, and hormonal responsiveness of polarized uterine epithelial cells, 154
Carbohydrate metabolism in preimplantation mouse embryos, 67–75
Cattle embryos. *See* Bovine embryo development
Cavitation. *See* Blastocoel formation in mouse
Cell-cell adhesion in implantation, 163–164
Cell cycles in bovine embryo development, 5–8
Cell differentiation. *See* Differentiation
Cell proliferation
   epidermal growth factor affecting, 126, 132
   insulin affecting, 115–116
Chick embryos, growth factors in, 19
Co-cultures
   of bovine oocytes
      with granulosa cells or cumulus cells, 4
      with oviduct epithelial cells, 4, 8
   of human embryos and fetal bovine uterine fibroblasts, 155
   of NR-6-NR cells with early mouse embryos, 60–61
Collagen IV in EHS matrix for uterine epithelial cells, 159
Complement protein C_3 secretion polarized uterine epithelial cells, 163
Crabtree effect, phosphate affecting, 84
Cumulus cells, bovine, 3
   co-culture with oocytes, 4
Cytochrome oxidase mRNA in mouse oocytes or preimplantation embryos, 35
Cytokeratins in polarized uterine epithelial cells, 159
Cytoplasmic maturation, and bovine embryo development, 3–5

DAPI staining for DNA analysis in embryos, 5, 31
Decidual production of insulin-like growth factor binding protein in primates, 137–147
Deoxyglucose, and hamster embryo development, 87
Differentiation
   and detection of low copy number mRNAs in mouse, 37
   of embryonal carcinoma cells, and production of growth factors, 53–63
   endometrial, affected by insulin-like growth factor, 145
   epidermal growth factor affecting, 126
   and gene activation from 2-cell stage in mouse, 29
   growth factors affecting, 12, 19–20
   and insulin receptor gene in early mouse embryo, 40
   trophectoderm, and gene expression in blastocoel formation, 97
DNA
   analysis in cell cycle 1 of bovine embryos, 5–6
   insulin affecting synthesis of, 110, 114, 122
cDNA clones for insulin-like growth factor binding protein, 141–142
Drosophila embryos, growth factors in, 19

Electron microscopy, high-resolution, of insulin binding to mouse embryo, 117, 118–119
Embryonal carcinoma cells
   detection of alkaline phosphatase and insulin receptor genes in, 37
   differentiation to parietal endoderm-like cells, 58, 61
   epidermal growth factor receptor in, 51
   growth factors produced by, 19, 53–63
Endocytosis of insulin, receptor-mediated, 118–121
Endoderm-like cells, parietal, differentiation from embryonal carcinoma cells, 58, 61
Endometriosis, and endometrial production of insulin-like growth factor binding protein, 143
Endometrium. *See* Uterus
Energy metabolism in preimplantation mouse embryos, 67–75
Entactin in EHS matrix for uterine epithelial cells, 159
Epidermal growth factor
   binding by embryonal carcinoma stem cells, 19
   binding by preimplantation mouse embryos, 18
   deficiency of, and abortions in mice, 126
   detection of mRNA by *in situ* hybridization, 128, 130

estrogen affecting regulation of, 20
functions of, 126
gene expression in uterus
  estrogen affecting, 20, 125, 132
  sialoadenectomy affecting, 125
  immunolocalization in uterus, 128–130
    and disappearance on day 5 of pregnancy, 128, 133
  in kidney of mouse, 126
  in plasma of mouse
    in pregnancy, 126
    sialoadenectomy affecting, 130, 131, 132
  and pregnancy in mouse, 125–133
  receptors for, 126
    expression on trophoblast giant cells in mouse, 47–48
    similarity to v-erb B oncogene, 126
    tyrosine kinase activity of, 50–51
  in submandibular gland of mouse, 126–130
  and pregnancy outcome, 126
Estrogen
  and epidermal growth factor gene expression in uterus, 20, 125, 132
  responsiveness of uterine epithelial cells in culture, 158
    polarity affecting, 163, 164–165
  and secretion of insulin-like growth factor protein, 140
Extracellular matrix in culture systems for uterine epithelial cells, 155, 158

Fatty acids as energy source for ova in rabbit, 74
Fibroblast cells, fetal bovine uterine, cultured with human embryos, 155
Fibroblast growth factor
  basic
    in humans, 57, 62
    in preimplantation mouse embryos, 12, 15, 17–20
  Kaposi sarcoma-type
    gene expression by embryonal carcinoma cells, 53–63
    in humans, 57, 62
    in preimplantation mouse embryos, 15, 17, 18
Fibronectin affecting trophoblast outgrowth, 59
Follicle size, and cytoplasmic maturation in cattle, 3–4
Follicle-stimulating hormone, and expansion of bovine cumulus cells, 3
c-Fos protein detection in trophoblast cell responses to serum stimulation, 48
Frog embryos
  growth factor in, 19, 20
  mesoderm induction in, growth factors affecting, 63

Gap junction communication, and blastocoel formation, 98
Gene expression
  activation of zygote genome affecting, 29
  for blastocoel formation in mouse, 97–105
  for epidermal growth factor in uterus
    estrogen affecting, 20, 125, 132
    sialoadenectomy affecting, 125
  for insulin receptors, 40–41
  for Kaposi sarcoma-type fibroblast growth factor in embryonal carcinoma cells, 53–63
Germ cells, primordial, alkaline phosphatase activity in mouse, 38
Glucose
  effects on spermatozoa, 83
  and hamster embryo development, 79–94
  homeostasis regulation by insulin, 145
  metabolism in preimplantation mouse embryos, 67–75
    and blockade of glycolysis, 70
  transport affected by insulin in mouse embryos, 110
Glutamic acid, and hamster embryo development, 83
Glutamine
  and energy metabolism in preimplantation embryos, 74, 75
  and hamster embryo development, 81, 90
Glycine, and hamster embryo development, 82, 83
Glycolysis aerobic, in preimplantation mouse embryos, 72–75
Gold-labelled insulin binding and uptake in mouse embryos, 112–113
  and internalization by coated pits, 119–121
  receptor-mediated, 117–120
Gonadotropins, and expansion of bovine cumulus cells, 3
Granulosa cells co-cultured with bovine oocytes, 4
Growth factors

embryonal carcinoma-derived, 53–63
epidermal. *See* Epidermal growth factor
insulin-like. *See* Insulin-like growth factor
maternal, and responses of mouse embryos, 20
in preimplantation mouse embryos, 11–21
  accumulation patterns of, 19–20
  function of, 20

Hamster embryo development, 79–94
  amino acids affecting, 81, 92
  and blastocyst attachment in cultures, 163
  deoxyglucose affecting, 87
  in eight-cell stage, 81, 85
    glucose and phosphate affecting, 90–91
  in four-cell stage, 81, 85
    phosphate affecting, 87–88, 91
  glucose inhibition of, 83–84
    phosphate affecting, 85
  in HECM-1 medium with 20 amino acids, 83, 89, 92
  in HECM-2 medium, 91, 92
  in microdrop cultures, 82–83
  phosphate affecting, 84–88
    in 4-cell embryos, 87–88, 91
    and synergism of phosphate and glucose, 91
  in TALP culture medium, 81–82, 91
  in T-PVA culture medium, 90
    with no energy substrates, 90
  in TLP-PVA culture medium, 82, 91
    with four amino acids, 87, 90, 92
    with twenty amino acids, 86, 92
  in transferred embryos, 81, 89–90
  in two-cell stage, 81, 85, 87
    swollen blastomeres in, 82, 93
Heparin sulfate proteoglycan
  in EHS matrix for uterine epithelial cells, 159
  secretion by polarized uterine epithelial cells, 162, 163
Hepatocyte energy metabolism in culture, 73
Hexokinase activity in mouse oocytes, 70
Histone mRNA
  accumulation in blastocoele formation, 103–104
  in mouse oocytes or preimplantation embryos, 35
Hoechst 3342 stain for mouse embryos, 112
Hst gene produced by human germ cell tumors, 54
Humans
  basic fibroblast growth factor in, 57, 62
  composition of culture medium affecting *in vitro* development, 155
  germ cell tumors producing transforming gene hst, 54
  insulin-like growth factor binding protein and pregnancy in, 137–147
  Kaposi sarcoma-type fibroblast growth factor in, 58, 62
Hypoglycemia, and serum levels of insulin-like growth factor binding protein, 142, 145

Immunocytochemical localization
  of epidermal growth factor in uterus, 128–130
  of insulin in mouse embryos, 113–114
  of insulin-like growth factor binding protein in pregnancy, 139–140
Implantation
  cell-cell adhesion in, 163–164
  in vitro blastocyst attachment on polarized uterine epithelia, 153–165
  peri-implantation period
    and changes in ovoreceptivity of uterus, 155–156, 162, 163
    and epidermal growth factor gene expression in mouse uterus, 125, 126
  role of decidual cells in, 139
  sialoadenectomy affecting, 130, 131
  and uterine microvasculature in rodents, 72
Insulin
  and glucose homeostasis regulation, 145
  gold-labelled, binding and uptake in mouse embryos, 112–113
    and internalization by coated pits, 119–121
    receptor-mediated, 117–120
  immunocytochemical localization of, 113–114
  in maternal reproductive tract, 113, 114
    as source of insulin in mouse blastocysts, 120–121
  metabolic effects of, 110, 114
  mitogenic effects of, 115–116
  and preimplantation mouse development, 109–122
  structure and function of, 110–111
  and synthesis of DNA and RNA, 114, 122

Index

Insulin-like growth factor
 action on endometrial/decidual cells
  affected by binding protein, 145
 binding site on binding protein, 141
 in preimplantation mouse embryos, 12, 15
 produced by embryonal carcinoma cells, 54
 receptors for, 139, 141, 144
  detection in early mouse embryos, 110
Insulin-like growth factor binding protein in primates, 137–147
 Arg-Gly-Asp tripeptide in, 141–142
 autocrine action of, 145–146
 diurnal variations in, 142
 cDNA clones for, 141–142
 in fetal liver, 142
 function of, 144–147
 growth hormone-dependent, 138
 growth hormone-independent, 138
 inhibitory and stimulatory effects of, 142, 145, 146
 localization of, 138, 139, 143, 146
 mitogenic action of, 140–141
 monoclonal antibodies to, 140
 paracrine action of, 146
 regulation of synthesis, 142–144
 serum levels in hypoglycemia, 142, 145
 synthesis and secretion of, 138–141
 systemic action of, 145
Insulin receptors
 gene detection
  in early mouse embryos, 39, 40–41, 110
  in embryonal carcinoma cells, 37
 morphometric analysis in mouse embryo, 117, 119
 structure and function of, 116–117
Int-2
 expression by embryonal carcinoma cells, 61–62
 in mouse embryos, 12
  compared to fibroblast growth factor family members, 57
Integrin-type receptors for insulin-like growth factor binding protein, 141, 144
Intracisternal A particles in mouse oocytes or preimplantation embryos, 35
Isoleucine, and hamster embryo development, 81, 90

Keratin sulfate proteoglycan secretion by polarized uterine epithelial cells, 162, 163
Kidney of mouse, epidermal growth factor in, 126

Lactate
 and hamster embryo development, 90, 91
 in preimplantation mouse embryos, 72–73
Lactate dehydrogenase activity in preimplantation mouse embryo, 75
Laminin in EHS matrix for uterine epithelial cells, 159
Liver
 energy metabolism in cultured cells, 73
 fetal, insulin-like growth factor binding protein in, 142

Maternal insulin in reproductive tract, 113, 114
 as source of insulin in mouse blastocyst, 120–121
Maternal to zygote genome control of mouse embryo development, 20, 28, 34, 35
Menstrual cycle
 localization of insulin-like growth factor binding protein in, 139, 143
 secretion of pregnancy-associated endometrial α_1 and α_2-globulins in luteal phase, 138
Mesoderm induction in frogs, growth factors in, 63
Metabolism
 insulin affecting, 110, 114
 in preimplantation mouse embryos, 67–75
Methionine
 and hamster embryo development, 81, 90
 uptake by polarized uterine epithelial cells, 159–161
Microdrop cultures, hamster embryo development in, 82–83
Milk, insulin-like growth factor binding protein in, 138
Mouse
 blastocoel formation in, 97–105
 blastocyst attachment and trophoblast outgrowth in cultures, 163
 blocks to development of embryos, 80
 energy metabolism in preimplantation embryo, 67–75
 epidermal growth factor and pregnancy in, 125–133

epidermal growth factor protein detection in embryo cells, 47–53
fibroblast growth factor-related growth factors produced by embryonal carcinoma cells and early embryos, 53–63
genomic activation and extended cell cycles in, 8
growth factors in preimplantation embryos, 11–21
insulin role in preimplantation embryo development, 109–122
requirements for embryo development, 81, 92
RNA and protein synthesis in preimplantation embryo development, 27–41
c-Myc induction affected by epidermal growth factor, 126

Na^+,K^+-ATPase, and timing of blastocoel formation in mouse, 97–105
Nerve growth factor binding by embryonal carcinoma stem cells, 19

Oocyte maturation, bovine, 2–3
Ouabain sensitivity of expanding blastocysts, 98
Oviducts
  co-culture of epithelial cells with bovine oocytes, 4, 8
  environment affecting embryo development, 83, 89, 93
  insulin localization in, 113
  rabbit, energy metabolism in, 74
Ovine embryos. *See* Sheep
Pasteur effect, phosphate affecting, 84
Paternal genome expression in 2-cell mouse embryo, 28
Peri-implantation period
  and changes in ovoreceptivity of uterus, 155–156, 162, 163
  and epidermal growth factor gene expression in mouse uterus, 125, 126
Phenylalanine, and hamster embryo development, 81, 90
Phosphate
  and hamster embryo development, 79–94
  turnover affected by epidermal growth factor, 126
Pig embryo, energy metabolism in, 70
Placental proteins 12 and 14, 138
Placental tissues, insulin-like growth factor in, 138
Platelet-derived growth factor
  in preimplantation mouse embryos, 12, 15, 17–20, 49
  produced by embryonal carcinoma cells, 54, 56
  and soft agar growth of NR-6-R cells, 61
Polar trophectoderm cells, insulin internalized by, 120
Polarized uterine epithelia, *in vitro* implantation on, 153–165
Polymerase chain reaction, and reverse transcription for mRNA phenotyping, 13, 17–18, 31–32, 37
Polyvinylalcohol in culture medium, and hamster embryo development, 82
Pregnancy
  ectopic, and production of insulin-like growth factor binding protein, 143
  and epidermal growth factor in mouse, 125–133
  and insulin-like growth factor binding protein in primates, 137–147
  sialoadenectomy affecting, in mice, 130, 131, 132
Pregnancy-associated endometrial α_1 and α_2-globulins, 138
Primates. *see also* Humans
  blocks to development in embryos, 80
  insulin-like growth factor binding protein in, 137–147
Progesterone
  and decidualization of stromal cells, 139
  responsiveness of uterine epithelial cells in culture, 158
  and secretion of insulin-like growth factor binding protein, 140, 143
  and transcription of insulin-like growth factor binding protein, 142
Prolactin production, factors affecting, 143, 144, 145–146
Prostaglandins
  epidermal growth factor affecting synthesis of, 126
  vectorial secretion of $PGF_2\alpha$ by polarized uterine epithelial cells, 162–163
Protein kinase activity of epidermal growth factor receptor, 50–51
Protein synthesis
  in bovine embryos
    in fourth cell cycle, 8
    in third cell cycle, 7
  changes during development of preim-

plantation mouse embryo, 27–41
  insulin affecting, 110
Proteoglycans secreted by polarized uterine epithelial cells, 162, 163
Pyruvate
  and hamster embryo development, 91
  metabolism in preimplantation mouse embryos, 67–75
Rabbit
  energy metabolism in embryos, 70
  fatty acids as energy source for ova, 74
  synthesis of Na^+,K^+-ATPase in embryo, 105
Rats
  blastocyst attachment and trophoblast outgrowth in cultures, 163–164
  blastocyst attachment to uterine epithelia, 154
  blocks to development in embryos, 80
Receptor(s)
  for epidermal growth factor, 126
    tyrosine kinase activity of, 50–51
  for insulin. See Insulin receptors
  for insulin-like growth factors, 139, 141, 144
Receptor-mediated insulin binding to mouse embryo, 117–120
Relaxin, and production of insulin-like growth factor binding protein, 144
Reproductive tract, maternal, insulin in, 113, 114
  as source of insulin in mouse embryo, 120–121
Reverse transcription, and polymerase chain reaction for mRNA phenotyping, 13, 17–18, 31–32, 37
Rhesus monkey production of insulin-like growth factor binding protein by decidual tissues, 144
RNA
  changes during development of preimplantation mouse embryo, 27–41
  insulin affecting synthesis of, 114, 122
  and meiosis resumption, 3
  small nuclear
    localization in mouse oocytes and preimplantation embryos, 35, 36
    in processing pre-mRNA, 28, 34
mRNA
  for catalytic subunit of Na^+,K^+-ATPase, detection in mouse embryos, 97–105
  epidermal growth factor, detection by *in situ* hybridization, 128, 130
  maternal, in transition to embryonic control in 2-cell mouse embryo, 28, 34, 35
phenotyping
  and detection of insulin receptors in mouse embryos, 110
  by reverse transcription-polymerase chain reaction, 13, 17–18, 31–32, 37
Sheep
  epithelial protein in uterine lumen in pregnancy, 146
  genomic activation and extended cell cycles in, 8
  protein synthesis in third cell cycle of embryos, 7
  trophoblast protein-1, 20
Sialoadenectomy
  effects on pregnancy in mice, 130, 131, 132
  and epidermal growth factor in uterus, 125, 128
  and plasma levels of epidermal growth factor, 130, 131, 132
  progestational, and abortions in mice, 126
Sm antigen localization in mouse oocytes and preimplantation embryos, 35, 36
Sodium pump, and timing of blastocoel formation in mouse, 97–105
Spermatozoa
  concentrations in cattle affecting embryo development, 4
  glucose affecting, 83
  synchronous sperm penetration and oocyte activation in cattle, 5
Steroid synthesis, epidermal growth factor affecting, 126
Stromelysin in preimplantation mouse embryo, 15
Submandibular gland of mouse, epidermal growth factor in, 126–130
Sulfur pertechnetate, and hamster embryo development, 88

Taurine, and hamster embryo development, 82, 83
Teratocarcinoma cells
  alkaline phosphatase activity in, 38

binding of epidermal growth factor, 48
growth factor produced by, 12
Tight junction formation in uterine epithelial cell cultures, 159
Tissue plasminogen activator mRNA in mouse oocytes or preimplantation embryos, 35
Transforming growth factors
  inducing soft agar growth of cell lines, 56
  in preimplantation mouse embryos, 12, 15, 17–20, 51
  produced by embryonal carcinoma cells, 54
Trophectoderm
  insulin in cells of, 119–120
  mural, Na^+,K^+-ATPase catalytic subunit detection in, 97–105
Trophoblast
  association with decidualized endometrium, 137, 138
  giant cells binding epidermal growth factor in mouse, 47–48
  insulin-like growth factor binding protein affecting, 140–141, 146
Tyrosine kinase activity
  of epidermal growth factor receptor, 50–51
  insulin affecting, 116

Uterus
  angiogenesis affected by embryonic growth factors, 20
  apical and basal secretions by polarized epithelial cells, 159–163
  contraction affected by epidermal growth factor, 126
  decidualized endometrium associated with trophoblast, 137, 138
  epidermal growth factor gene expression in, 125, 132
  growth promoting effects of insulin-like growth factor in endometrium, 145
  interactions of endometrium with blastocyst, 156
  interactions of epithelial and stromal cells, 162
  in vitro blastocyst attachment on polarized epithelia, 153–165
  microvasculature during implantation in rodents, 72
  peri-implantation changes in ovoreceptivity, 155–156, 162, 163
  requirements for epithelial cell cultures
    matrix in, 159
    permeable culture surface in, 159
  secretion of insulin-like growth factor binding protein in primates, 137–147
Uvomorulin in polarized uterine epithelial cells, 159

Xenopus embryos
  cell cycles in, 12
  genomic activation and protein synthesis in, 8
  growth factors in, 19, 20

Zona pellucida mRNA in mouse oocytes or preimplantation embryos, 35

MAY 1 5 1990